Contents

Getting the most from this book

Mathematics is not only a beautiful and exciting subject in its own right but also one that underpins many other branches of learning. It is consequently fundamental to our national wellbeing.

This book covers the compulsory core content of Year 1/AS Further Mathematics and so provides material for the first of the two years of Advanced Level Further Mathematics study. The requirements of the compulsory core content for the second year are met in a second book, while Year 1 and Year 2 optional applied content is covered in the Mechanics and Statistics books.

Between 2014 and 2016 A Level Mathematics and Further Mathematics were substantially revised, for first teaching in 2017. Major changes included increased emphasis on:

■ Problem solving

■ Mathematical proof

■ Use of ICT

■ Modelling

■ Working with large data sets in statistics.

This book embraces these ideas. A large number of exercise questions involve elements of problem solving. The ideas of mathematical proof, rigorous logical argument and mathematical modelling are also included in suitable exercise questions throughout the book.

The use of technology, including graphing software, spreadsheets and high specification calculators, is encouraged wherever possible, for example in the Activities used to introduce some of the topics. In particular, readers are expected to have access to a calculator which handles matrices up to order 3×3. Places where ICT can be used are highlighted by a (T) icon. Margin boxes highlight situations where the use of technology – such as graphical calculators or graphing software – can be used to further explore a particular topic.

Throughout the book the emphasis is on understanding and interpretation rather than mere routine calculations, but the various exercises do nonetheless provide plenty of scope for practising basic techniques. The exercise questions are split into three bands. Band 1 questions are designed to reinforce basic understanding; Band 2 questions are broadly typical of what might be expected in an examination; Band 3 questions explore around the topic and some of them are rather more demanding. In addition, extensive online support tailored to the Edexcel specification, including further questions, is available by subscription to MEI's Integral website, integralmaths.org.

In addition to the exercise questions, there are two sets of Practice questions, covering groups of chapters. These include identified questions requiring problem solving (PS), mathematical proof (MP), use of ICT (T) and modelling (M).

This book is written on the assumption that readers are studying or have studied AS Mathematics. It can be studied alongside the Year 1/AS Mathematics book, or after studying AS or A Level Mathematics. There are places where the work depends on knowledge from earlier in the book or in the Year 1/AS Mathematics book and this is flagged up in the Prior knowledge boxes. This should be seen as an invitation to those who have problems with the particular topic to revisit it. At the end of each chapter there is a list of key points covered as well as a summary of the new knowledge (learning outcomes) that readers should have gained.

Although a general knowledge of A Level Mathematics beyond AS Level is not required, there are two small topics from Year 2 of A Level Mathematics that are needed in the study of the material in this book. These are radians (needed in the work on the argument of a complex number) and the compound angle formulae, which are helpful in understanding the multiplication and division of complex numbers in modulus-argument form. These two topics are introduced briefly at the back of the book, for the benefit of readers who have not yet studied Year 2 of A Level Mathematics.

Two common features of the book are Activities and Discussion points. These serve rather different purposes. The Activities are designed to help readers get into the thought processes of the new work that they are about to meet; having done an Activity, what follows will seem much easier. The Discussion points invite readers to talk about particular points with their fellow students and their teacher and so enhance their understanding. Another feature is a Caution icon ❗, highlighting points where it is easy to go wrong.

Answers to all exercise questions and practice questions are provided at the back of the book, and also online at www.hoddereducation.co.uk/EdexcelFurtherMathsYear1

Catherine Berry

Roger Porkess

Acknowledgements

The Publishers would like to thank the following for permission to reproduce copyright material.

Practice questions have been provided by MEI (p. 84–85 and p. 197–198).

Photo credits

p.1 © ironstu – iStock via Thinkstock/Getty Images; **p.35** © Markus Mainka/Fotolia; **p.38** © Wellcom Images via Wikipedia (https://creativecommons.org/licenses/by/4.0/); **p.48** © Dusso Janladde via Wikipedia Commons (https://en.wikipedia.org/wiki/GNU_Free_Documentation_License); **p.58** Public Domain; **p.66 (top)** © Charles Brutlag – 123RF; **p.66 (bottom)** © oriontrail – iStock via Thinkstock; **p.86** © Creative-Family – iStock via Thinkstock; **p.96 (top)** © Photodisc/Getty Images/ Business & Industry 1; **p.96 (bottom)** © Wolfgang Beyer (Wikipedia Commons, https://creativecommons.org/licenses/by-sa/3.0/deed.en); **p.122** © marcel/ Fotolia; **p.147** © lesley marlor/Fotolia.

EDEXCEL A LEVEL

SET TEXT BOOK

FURTHER MATHEMATICS

For Core Year 1 and AS

1

BEN SPARKS, CLAIRE BALDWIN
SERIES EDITORS

ROGER PORKESS AND CATHERINE BERRY
CONSULTANT EDITORS

KEITH PLEDGER AND JAN DANGERFIELD

HODDER EDUCATION
LEARN MORE

In order to ensure that this resource offers high-quality support for the associated Pearson qualification, it has been through a review process by the awarding body. This process confirms that this resource fully covers the teaching and learning content of the specification or part of a specification at which it is aimed. It also confirms that it demonstrates an appropriate balance between the development of subject skills, knowledge and understanding, in addition to preparation for assessment.

Endorsement does not cover any guidance on assessment activities or processes (e.g. practice questions or advice on how to answer assessment questions), included in the resource nor does it prescribe any particular approach to the teaching or delivery of a related course.

While the publishers have made every attempt to ensure that advice on the qualification and its assessment is accurate, the official specification and associated assessment guidance materials are the only authoritative source of information and should always be referred to for definitive guidance.

Pearson examiners have not contributed to any sections in this resource relevant to examination papers for which they have responsibility.

Examiners will not use endorsed resources as a source of material for any assessment set by Pearson.

Endorsement of a resource does not mean that the resource is required to achieve this Pearson qualification, nor does it mean that it is the only suitable material available to support the qualification, and any resource lists produced by the awarding body shall include this and other appropriate resources.

Hachette UK's policy is to use papers that are natural, renewable and recyclable products and made from wood grown in sustainable forests. The logging and manufacturing processes are expected to conform to the environmental regulations of the country of origin.

Orders: please contact Bookpoint Ltd, 130 Park Drive, Milton Park, Abingdon, Oxon OX14 4SE. Telephone: (44) 01235 827720. Fax: (44) 01235 400454. Email: education@bookpoint.co.uk. Lines are open from 9 a.m. to 5 p.m., Monday to Saturday, with a 24-hour message answering service. You can also order through our website: www.hoddereducation.co.uk

ISBN: 978 1471 860218

© Jan Dangerfield, Roger Porkess, Ben Sparks and MEI 2017

First published in 2017 by

Hodder Education
An Hachette UK Company
Carmelite House
50 Victoria Embankment
London EC4Y 0DZ
www.hoddereducation.co.uk

Impression number   10 9 8 7 6 5 4 3 2 1
Year   2021   2020   2019   2018   2017

Cover photo © delcreations/123RF.com

Typeset in Bembo Std, 11/13 pts. by Aptara, Inc.

Printed in Italy

A catalogue record for this title is available from the British Library.

1 Matrices and transformations

As for everything else, so for a mathematical theory – beauty can be perceived but not explained.

Arthur Cayley 1883

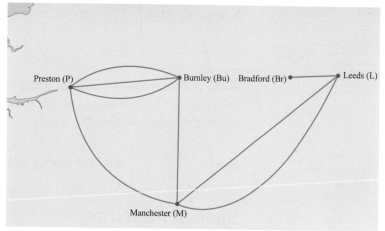

Figure 1.1 Illustration of some major roads and motorways joining some towns and cities in the north of England.

Discussion point

→ How many direct routes (without going through any other town) are there from Preston to Burnley? What about Manchester to Leeds? Preston to Manchester? Burnley to Leeds?

1 Matrices

You can represent the number of direct routes between each pair of towns (shown in Figure 1.1) in an array of numbers like this:

	Br	Bu	L	M	P
Br	0	0	1	0	0
Bu	0	0	0	1	3
L	1	0	0	2	0
M	0	1	2	0	1
P	0	3	0	1	0

This array is called a matrix (the plural is matrices) and is usually written inside curved brackets.

$$\begin{pmatrix} 0 & 0 & 1 & 0 & 0 \\ 0 & 0 & 0 & 1 & 3 \\ 1 & 0 & 0 & 2 & 0 \\ 0 & 1 & 2 & 0 & 1 \\ 0 & 3 & 0 & 1 & 0 \end{pmatrix}$$

It is usual to represent matrices by capital letters, often in bold print.

A matrix consists of rows and columns, and the entries in the various cells are known as **elements**.

The matrix $\mathbf{M} = \begin{pmatrix} 0 & 0 & 1 & 0 & 0 \\ 0 & 0 & 0 & 1 & 3 \\ 1 & 0 & 0 & 2 & 0 \\ 0 & 1 & 2 & 0 & 1 \\ 0 & 3 & 0 & 1 & 0 \end{pmatrix}$ representing the routes between the

towns and cities has 25 elements, arranged in five rows and five columns. $\mathbf{M}$ is described as a 5×5 matrix, and this is the **order** of the matrix. You state the number of rows first, then the number of columns. So, for example, the matrix

$\mathbf{A} = \begin{pmatrix} 3 & -1 & 4 \\ 2 & 0 & 5 \end{pmatrix}$ is a 2×3 matrix and $\mathbf{B} = \begin{pmatrix} 4 & -4 \\ 3 & 4 \\ 0 & -2 \end{pmatrix}$ is a 3×2 matrix.

Special matrices

Some matrices are described by special names which relate to the number of rows and columns or the nature of the elements.

Matrices such as $\begin{pmatrix} 4 & 2 \\ 1 & 0 \end{pmatrix}$ and $\begin{pmatrix} 3 & 5 & 1 \\ 2 & 0 & -4 \\ 1 & 7 & 3 \end{pmatrix}$ which have the same number of

rows as columns are called **square matrices**.

The matrix $\begin{pmatrix} 1 & 0 \\ 0 & 1 \end{pmatrix}$ is called the 2×2 **identity matrix** or **unit matrix**, and

similarly $\begin{pmatrix} 1 & 0 & 0 \\ 0 & 1 & 0 \\ 0 & 0 & 1 \end{pmatrix}$ is called the 3×3 identity matrix. Identity matrices must

be square, and are usually denoted by **I**. An identity matrix consists of 1's on the leading diagonal (the diagonal from top left to bottom right) and 0's everywhere else.

The matrix $\mathbf{O} = \begin{pmatrix} 0 & 0 \\ 0 & 0 \end{pmatrix}$ is called the 2×2 **zero matrix**. Zero matrices can be of any order.

Two matrices are said to be **equal** if and only if they have the same order and each element in one matrix is equal to the corresponding element in the other matrix. So, for example, the matrices **A** and **D** below are equal, but **B** and **C** are not equal to any of the other matrices.

$$\mathbf{A} = \begin{pmatrix} 1 & 3 \\ 2 & 4 \end{pmatrix} \qquad \mathbf{B} = \begin{pmatrix} 1 & 2 \\ 3 & 4 \end{pmatrix} \qquad \mathbf{C} = \begin{pmatrix} 1 & 3 & 0 \\ 2 & 4 & 0 \end{pmatrix} \qquad \mathbf{D} = \begin{pmatrix} 1 & 3 \\ 2 & 4 \end{pmatrix}$$

Working with matrices

Matrices can be added or subtracted if they are of the same order.

$$\begin{pmatrix} 2 & 4 & 0 \\ -1 & 3 & 5 \end{pmatrix} + \begin{pmatrix} 1 & -1 & 4 \\ 2 & 0 & -5 \end{pmatrix} = \begin{pmatrix} 3 & 3 & 4 \\ 1 & 3 & 0 \end{pmatrix}$$
Add the elements in corresponding positions.

$$\begin{pmatrix} 2 & -3 \\ 4 & 1 \end{pmatrix} - \begin{pmatrix} 7 & -3 \\ -1 & 2 \end{pmatrix} = \begin{pmatrix} -5 & 0 \\ 5 & -1 \end{pmatrix}$$
Subtract the elements in corresponding positions.

But $\begin{pmatrix} 2 & 4 & 0 \\ -1 & 3 & 5 \end{pmatrix} + \begin{pmatrix} 2 & -3 \\ 4 & 1 \end{pmatrix}$ cannot be evaluated because the matrices are

not of the same order. These matrices are **non-conformable** for addition.

You can also multiply a matrix by a **scalar** number:

$$2 \begin{pmatrix} 3 & -4 \\ 0 & 6 \end{pmatrix} = \begin{pmatrix} 6 & -8 \\ 0 & 12 \end{pmatrix}$$
Multiply each of the elements by 2.

🖥 TECHNOLOGY

You can use a calculator to add and subtract matrices of the same order and to multiply a matrix by a number. For your calculator, find out:
- the method for inputting matrices
- how to add and subtract matrices
- how to multiply a matrix by a number for matrices of varying sizes.

Associativity and commutativity

When working with numbers the properties of **associativity** and **commutativity** are often used.

Associativity

Addition of numbers is **associative**.

$$(3 + 5) + 8 = 3 + (5 + 8)$$

When you add numbers, it does not matter how the numbers are grouped, the answer will be the same.

Discussion points

→ Give examples to show that subtraction of numbers is not commutative or associative.

→ Are matrix addition and matrix subtraction associative and/or commutative?

Commutativity

Addition of numbers is **commutative**.

$$4 + 5 = 5 + 4$$

When you add numbers, the order of the numbers can be reversed and the answer will still be the same.

Exercise 1.1

① Write down the order of these matrices.

(i) $\begin{pmatrix} 2 & 4 \\ 6 & 0 \\ -3 & 7 \end{pmatrix}$
(ii) $\begin{pmatrix} 0 & 8 & 4 \\ -2 & -3 & 1 \\ 5 & 3 & -2 \end{pmatrix}$
(iii) $\begin{pmatrix} 7 & -3 \end{pmatrix}$
(iv) $\begin{pmatrix} 1 \\ 2 \\ 3 \\ 4 \\ 5 \end{pmatrix}$

(v) $\begin{pmatrix} 2 & -6 & 4 & 9 \\ 5 & 10 & 11 & -4 \end{pmatrix}$
(vi) $\begin{pmatrix} 8 & 5 \\ -2 & 0 \\ 3 & -9 \end{pmatrix}$

② For the matrices

$$\mathbf{A} = \begin{pmatrix} 2 & -3 \\ 0 & 4 \end{pmatrix} \quad \mathbf{B} = \begin{pmatrix} 7 & -3 \\ 1 & 4 \end{pmatrix} \quad \mathbf{C} = \begin{pmatrix} 3 & 5 & -9 \\ 2 & 1 & 4 \end{pmatrix} \quad \mathbf{D} = \begin{pmatrix} 0 & -4 & 5 \\ 2 & 1 & 8 \end{pmatrix}$$

$$\mathbf{E} = \begin{pmatrix} -3 & 5 \\ -2 & 7 \end{pmatrix} \quad \mathbf{F} = \begin{pmatrix} 1 \\ 3 \\ 5 \end{pmatrix}$$

find, where possible

(i) $\mathbf{A} - \mathbf{E}$
(ii) $\mathbf{C} + \mathbf{D}$
(iii) $\mathbf{E} + \mathbf{A} - \mathbf{B}$
(iv) $\mathbf{F} + \mathbf{D}$
(v) $\mathbf{D} - \mathbf{C}$
(vi) $4\mathbf{F}$
(vii) $3\mathbf{C} + 2\mathbf{D}$
(viii) $\mathbf{B} + 2\mathbf{F}$
(ix) $\mathbf{E} - (2\mathbf{B} - \mathbf{A})$

③ The diagram in Figure 1.2 shows the number of direct flights on one day offered by an airline between cities P, Q, R and S.

The same information is also given in the partly completed matrix **X**.

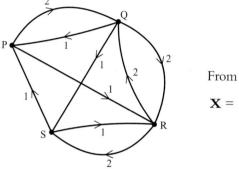

$$\mathbf{X} = \begin{array}{c} \\ \text{From} \begin{array}{c} \text{P} \\ \text{Q} \\ \text{R} \\ \text{S} \end{array} \end{array} \begin{array}{c} \text{To} \\ \begin{array}{cccc} \text{P} & \text{Q} & \text{R} & \text{S} \end{array} \\ \begin{pmatrix} 0 & 2 & 1 & 0 \\ 1 & & & \\ & & & \\ & & & \end{pmatrix} \end{array}$$

Figure 1.2

(i) Copy and complete the matrix **X**.

A second airline also offers flights between these four cities. The following matrix represents the total number of direct flights offered by the two airlines.

$$\begin{pmatrix} 0 & 2 & 3 & 2 \\ 2 & 0 & 2 & 1 \\ 2 & 2 & 0 & 3 \\ 1 & 0 & 3 & 0 \end{pmatrix}$$

(ii) Find the matrix **Y** representing the flights offered by the second airline.

(iii) Draw a diagram similar to the one in Figure 1.2, showing the flights offered by the second airline.

④ Find the values of w, x, y and z such that

$$\begin{pmatrix} 3 & w \\ -1 & 4 \end{pmatrix} + x \begin{pmatrix} 2 & -1 \\ y & z \end{pmatrix} = \begin{pmatrix} -9 & 8 \\ 11 & -8 \end{pmatrix}.$$

⑤ Find the possible values of p and q such that

$$\begin{pmatrix} p^2 & -3 \\ 2 & 9 \end{pmatrix} - \begin{pmatrix} 5p & -2 \\ -7 & q^2 \end{pmatrix} = \begin{pmatrix} 6 & -1 \\ 9 & 4 \end{pmatrix}.$$

⑥ Four local football teams took part in a competition in which they each played each other twice, once at home and once away. Figure 1.3 shows the results matrix after half of the games had been played.

	Win	Draw	Lose	Goals for	Goals against
City	2	1	0	6	3
Rangers	0	0	3	2	8
Town	2	0	1	4	3
United	1	1	1	5	3

Figure 1.3

(i) The results of the next three matches are as follows:

City 2   Rangers 0

Town 3   United 3

City 2   Town 4

Find the results matrix for these three matches and hence find the complete results matrix for all the matches so far.

(ii) Here is the complete results matrix for the whole competition.

$$\begin{pmatrix} 4 & 1 & 1 & 12 & 8 \\ 1 & 1 & 4 & 5 & 12 \\ 3 & 1 & 2 & 12 & 10 \\ 1 & 3 & 2 & 10 & 9 \end{pmatrix}$$

Find the results matrix for the last three matches (City vs United, Rangers vs Town and Rangers vs United) and deduce the result of each of these three matches.

⑦ A mail-order clothing company stocks a jacket in three different sizes and four different colours.

The matrix $\mathbf{P} = \begin{pmatrix} 17 & 8 & 10 & 15 \\ 6 & 12 & 19 & 3 \\ 24 & 10 & 11 & 6 \end{pmatrix}$ represents the number of jackets in

stock at the start of one week.

The matrix $\mathbf{Q} = \begin{pmatrix} 2 & 5 & 3 & 0 \\ 1 & 3 & 4 & 6 \\ 5 & 0 & 2 & 3 \end{pmatrix}$ represents the number of orders for

jackets received during the week.

(i) Find the matrix $\mathbf{P} - \mathbf{Q}$.

What does this matrix represent? What does the negative element in the matrix mean?

A delivery of jackets is received from the manufacturers during the week.

The matrix $\mathbf{R} = \begin{pmatrix} 5 & 10 & 10 & 5 \\ 10 & 10 & 5 & 15 \\ 0 & 0 & 5 & 5 \end{pmatrix}$ shows the number of jackets received.

(ii) Find the matrix which represents the number of jackets in stock at the end of the week after all the orders have been dispatched.

(iii) Assuming that this week is typical, find the matrix which represents sales of jackets over a six-week period. How realistic is this assumption?

2 Multiplication of matrices

When you multiply two matrices you do not just multiply corresponding terms. Instead you follow a slightly more complicated procedure. The following example will help you to understand the rationale for the way it is done.

There are four ways of scoring points in rugby: a try (five points), a conversion (two points), a penalty (three points) and a drop goal (three points). In a match Tonga scored three tries, one conversion, two penalties and one drop goal.

So their score was

$3 \times 5 + 1 \times 2 + 2 \times 3 + 1 \times 3 = 26.$

You can write this information using matrices. The tries, conversions, penalties and drop goals that Tonga scored are written as the 1×4 row matrix $(3\ \ 1\ \ 2\ \ 1)$ and the points for the different methods of scoring as the 4×1 column matrix

$$\begin{pmatrix} 5 \\ 2 \\ 3 \\ 3 \end{pmatrix}.$$

These are combined to give the 1×1 matrix $(3 \times 5 + 1 \times 2 + 2 \times 3 + 1 \times 3) = (26).$

Combining matrices in this way is called **matrix multiplication** and this

example is written as $(3\ \ 1\ \ 2\ \ 1)\begin{pmatrix} 5 \\ 2 \\ 3 \\ 3 \end{pmatrix} = (26).$

The use of matrices can be extended to include the points scored by the other team, Japan. They scored two tries, two conversions, four penalties and one drop goal. This information can be written together with Tonga's scores as a 2×4 matrix, with one row for Tonga and the other for Japan. The multiplication is then written as

$$\begin{pmatrix} 3 & 1 & 2 & 1 \\ 2 & 2 & 4 & 1 \end{pmatrix} \begin{pmatrix} 5 \\ 2 \\ 3 \\ 3 \end{pmatrix} = \begin{pmatrix} 26 \\ 29 \end{pmatrix}.$$

So Japan scored 29 points and won the match.

This example shows you two important points about matrix multiplication. Look at the orders of the matrices involved.

The two 'middle' numbers, in this case 4, must be the same for it to be possible to multiply two matrices. If two matrices can be multiplied, they are conformable for multiplication.

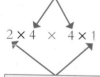

$2 \times 4\ \ \times\ \ 4 \times 1$

The two 'outside' numbers give you the order of the product matrix, in this case 2×1.

You can see from the previous example that multiplying matrices involves multiplying each element in a row of the left-hand matrix by each element in a column of the right-hand matrix and then adding these products.

Example 1.1

Find $\begin{pmatrix} 10 & 3 \\ -2 & 7 \end{pmatrix}\begin{pmatrix} 5 \\ 2 \end{pmatrix}$.

Solution

The product will have order 2×1.

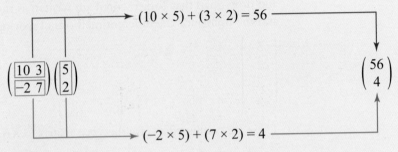

Figure 1.4

Example 1.2

Find $\begin{pmatrix} 1 & 3 \\ -2 & 5 \end{pmatrix}\begin{pmatrix} 4 & 3 & 0 \\ -2 & -3 & 1 \end{pmatrix}$.

Solution

The order of this product is 2×3.

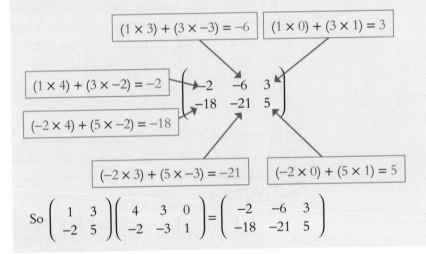

So $\begin{pmatrix} 1 & 3 \\ -2 & 5 \end{pmatrix}\begin{pmatrix} 4 & 3 & 0 \\ -2 & -3 & 1 \end{pmatrix} = \begin{pmatrix} -2 & -6 & 3 \\ -18 & -21 & 5 \end{pmatrix}$

Discussion point

➔ If $\mathbf{A} = \begin{pmatrix} 1 & 3 & 5 \\ -2 & 4 & 1 \\ 0 & 3 & 7 \end{pmatrix}$, $\mathbf{B} = \begin{pmatrix} 8 & -1 \\ -2 & 3 \\ 4 & 0 \end{pmatrix}$ and $\mathbf{C} = \begin{pmatrix} 5 & 0 \\ 3 & -4 \end{pmatrix}$

which of the products **AB**, **BA**, **AC**, **CA**, **BC** and **CB** exist?

Example 1.3

Find $\begin{pmatrix} 3 & 2 \\ -1 & 4 \end{pmatrix}\begin{pmatrix} 1 & 0 \\ 0 & 1 \end{pmatrix}$.

What do you notice?

Solution

The order of this product is 2×2.

$$\begin{pmatrix} 3 & 2 \\ -1 & 4 \end{pmatrix}\begin{pmatrix} 1 & 0 \\ 0 & 1 \end{pmatrix} = \begin{pmatrix} 3 & 2 \\ -1 & 4 \end{pmatrix}$$

$(3 \times 1) + (2 \times 0) = 3$

$(3 \times 0) + (2 \times 1) = 2$

$(-1 \times 0) + (4 \times 1) = 4$

$(-1 \times 1) + (4 \times 0) = -1$

Multiplying a matrix by the identity matrix has no effect.

Properties of matrix multiplication

In this section you will look at whether matrix multiplication is:

■ commutative

■ associative.

On page 4 you saw that for numbers, addition is both associative and commutative. Multiplication is also both associative and commutative. For example:

$$(3 \times 4) \times 5 = 3 \times (4 \times 5)$$

and

$$3 \times 4 = 4 \times 3$$

ACTIVITY 1.1

Using $\mathbf{A} = \begin{pmatrix} 2 & -1 \\ 3 & 4 \end{pmatrix}$ and $\mathbf{B} = \begin{pmatrix} -4 & 0 \\ -2 & 1 \end{pmatrix}$ find the products $\mathbf{AB}$ and $\mathbf{BA}$ and

hence comment on whether or not matrix multiplication is commutative.
Find a different pair of matrices, $\mathbf{C}$ and $\mathbf{D}$, such that $\mathbf{CD} = \mathbf{DC}$.

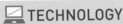
ACTIVITY 1.2

Using $\mathbf{A} = \begin{pmatrix} 2 & -1 \\ 3 & 4 \end{pmatrix}$, $\mathbf{B} = \begin{pmatrix} -4 & 0 \\ -2 & 1 \end{pmatrix}$ and $\mathbf{C} = \begin{pmatrix} 1 & 2 \\ 2 & 3 \end{pmatrix}$, find the matrix products:

(i)   **AB**
(ii)   **BC**
(iii)   **(AB)C**
(iv)   **A(BC)**

Does your answer suggest that matrix multiplication is associative?

Is this true for all 2×2 matrices? How can you prove your answer?

Exercise 1.2

In this exercise, do not use a calculator unless asked to. A calculator can be used for checking answers.

① Write down the orders of these matrices.

(i)   (a) $\mathbf{A} = \begin{pmatrix} 3 & 4 & -1 \\ 0 & 2 & 3 \\ 1 & 5 & 0 \end{pmatrix}$ (b) $\mathbf{B} = \begin{pmatrix} 2 & 3 & 6 \end{pmatrix}$

(c) $\mathbf{C} = \begin{pmatrix} 4 & 9 & 2 \\ 1 & -3 & 0 \end{pmatrix}$ (d) $\mathbf{D} = \begin{pmatrix} 0 & 2 & 4 & 2 \\ 0 & -3 & -8 & 1 \end{pmatrix}$

(e) $\mathbf{E} = \begin{pmatrix} 3 \\ 6 \end{pmatrix}$ (f) $\mathbf{F} = \begin{pmatrix} 2 & 5 & 0 & -4 & 1 \\ -3 & 9 & -3 & 2 & 2 \\ 1 & 0 & 0 & 10 & 4 \end{pmatrix}$

(ii)   Which of the following matrix products can be found? For those that can state the order of the matrix product.
(a) **AE**   (b) **AF**   (c) **FA**   (d) **CA**   (e) **DC**

② Calculate these products.

(i) $\begin{pmatrix} 3 & 0 \\ 5 & -1 \end{pmatrix} \begin{pmatrix} 7 & 2 \\ 4 & -3 \end{pmatrix}$

(ii) $\begin{pmatrix} 2 & -3 & 5 \end{pmatrix} \begin{pmatrix} 0 & 2 \\ 5 & 8 \\ -3 & 1 \end{pmatrix}$

(iii) $\begin{pmatrix} 2 & 5 & -1 & 0 \\ 3 & 6 & 4 & -3 \end{pmatrix} \begin{pmatrix} 1 \\ -9 \\ 11 \\ -2 \end{pmatrix}$

Check your answers using the matrix function on a calculator if possible.

③ Using the matrices $\mathbf{A} = \begin{pmatrix} 5 & 9 \\ -2 & 7 \end{pmatrix}$ and $\mathbf{B} = \begin{pmatrix} -3 & 5 \\ 2 & -9 \end{pmatrix}$, confirm that matrix multiplication is not commutative.

④ For the matrices

$$\mathbf{A} = \begin{pmatrix} 3 & 1 \\ 2 & 4 \end{pmatrix} \quad \mathbf{B} = \begin{pmatrix} -3 & 7 \\ 2 & 5 \end{pmatrix} \quad \mathbf{C} = \begin{pmatrix} 2 & 3 & 4 \\ 5 & 7 & 1 \end{pmatrix}$$

$$\mathbf{D} = \begin{pmatrix} 3 & 4 \\ 7 & 0 \\ 1 & -2 \end{pmatrix} \quad \mathbf{E} = \begin{pmatrix} 4 & 7 \\ 3 & -2 \\ 1 & 5 \end{pmatrix} \quad \mathbf{F} = \begin{pmatrix} 3 & 7 & -5 \\ 2 & 6 & 0 \\ -1 & 4 & 8 \end{pmatrix}$$

calculate, where possible, the following:

(i) **AB** (ii) **BA** (iii) **CD** (iv) **DC** (v) **EF** (vi) **FE**

⑤ Using the matrix function on a calculator, find $\mathbf{M}^4$ for the matrix

$$\mathbf{M} = \begin{pmatrix} 2 & 0 & -1 \\ 3 & 1 & 2 \\ -1 & 4 & 3 \end{pmatrix}.$$

> **Note**
> ---
> $\mathbf{M}^4$ means $\mathbf{M} \times \mathbf{M} \times \mathbf{M} \times \mathbf{M}$

⑥ $\mathbf{A} = \begin{pmatrix} x & 3 \\ 0 & -1 \end{pmatrix}$ $\mathbf{B} = \begin{pmatrix} 2x & 0 \\ 4 & -3 \end{pmatrix}$:

(i) Find the matrix product **AB** in terms of x.

(ii) If $\mathbf{AB} = \begin{pmatrix} 10x & -9 \\ -4 & 3 \end{pmatrix}$, find the possible values of x.

(iii) Find the possible matrix products **BA**.

⑦ (i) For the matrix $\mathbf{A} = \begin{pmatrix} 2 & 1 \\ 0 & 1 \end{pmatrix}$, find

(a) $\mathbf{A}^2$

(b) $\mathbf{A}^3$

(c) $\mathbf{A}^4$

(ii) Suggest a general form for the matrix $\mathbf{A}^n$ in terms of n.

(iii) Verify your answer by finding $\mathbf{A}^{10}$ on your calculator and confirming it gives the same answer as using (iv).

⑧ The map in Figure 1.5 below shows the bus routes in a holiday area. Lines represent routes that run each way between the resorts. Arrows indicated one-way scenic routes.

M is the partly completed 4 × 4 matrix which shows the number of direct routes between the various resorts.

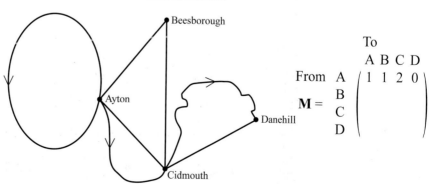

Figure 1.5

(i) Copy and complete the matrix **M**.

(ii) Calculate **M²** and explain what information it contains.

(iii) What information would **M³** contain?

⑨ $\mathbf{A} = \begin{pmatrix} 4 & x & 0 \\ 2 & -3 & 1 \end{pmatrix}$ $\mathbf{B} = \begin{pmatrix} 2 & -5 \\ 4 & x \\ x & 7 \end{pmatrix}$:

(i) Find the product **AB** in terms of x.

A symmetric matrix is one in which the entries are symmetrical about the

leading diagonal, for example $\begin{pmatrix} 2 & 5 \\ 5 & 0 \end{pmatrix}$ and $\begin{pmatrix} 3 & 4 & -6 \\ 4 & 2 & 5 \\ -6 & 5 & 1 \end{pmatrix}$.

(ii) Given that the matrix **AB** is symmetric, find the possible values of x.

(iii) Write down the possible matrices **AB**.

⑩ The matrix **A**, in Figure 1.6, shows the number of sales of five flavours of ice cream: Vanilla(V), Strawberry(S), Chocolate(C), Toffee(T) and Banana(B), from an ice cream shop on each of Wednesday(W), Thursday(Th), Friday(F) and Saturday(Sa) during one week.

$$\mathbf{A} = \begin{array}{c} \\ W \\ Th \\ F \\ Sa \end{array} \begin{array}{c} \begin{array}{ccccc} V & S & C & T & B \end{array} \\ \begin{pmatrix} 63 & 49 & 55 & 44 & 18 \\ 58 & 52 & 66 & 29 & 26 \\ 77 & 41 & 81 & 39 & 25 \\ 101 & 57 & 68 & 63 & 45 \end{pmatrix} \end{array}$$

Figure 1.6

(i) Find a matrix **D** such that the product **DA** shows the total number of sales of each flavour of ice cream during the four-day period and find the product **DA**.

(ii) Find a matrix **F** such that the product **AF** gives the total number of ice cream sales each day during the four-day period and find the product **AF**.

The Vanilla and Banana ice creams are served with strawberry sauce; the other three ice creams are served with chocolate sprinkles.

(iii) Find two matrices, **S** and **C**, such that the product **DAS** gives the total number of servings of strawberry sauce needed and the product **DAC** gives the total number of servings of sprinkles needed during the four-day period. Find the matrices **DAS** and **DAC**.

The price of Vanilla and Strawberry ice creams is 95p, Chocolate ice creams cost £1.05 and Toffee and Banana ice creams cost £1.15 each.

(iv) Using only matrix multiplication, find a way of calculating the total cost of all of the ice creams sold during the four-day period.

⑪ Figure 1.7 shows the start of the plaiting process for producing a leather bracelet from three leather strands a, b and c.

The process has only two steps, repeated alternately:

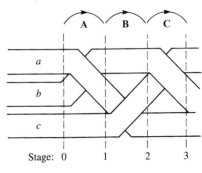

Figure 1.7

Step 1: cross the top strand over the middle strand

Step 2: cross the middle strand under the bottom strand.

At the start of the plaiting process, Stage 0, the order of the strands is given by $\mathbf{S}_0 = \begin{pmatrix} a \\ b \\ c \end{pmatrix}$.

(i) Show that pre-multiplying $\mathbf{S}_0$ by the matrix $\mathbf{A} = \begin{pmatrix} 0 & 1 & 0 \\ 1 & 0 & 0 \\ 0 & 0 & 1 \end{pmatrix}$ gives $\mathbf{S}_1$, the matrix which represents the order of the strands at Stage 1.

(ii) Find the 3×3 matrix $\mathbf{B}$ which represents the transition from Stage 1 to Stage 2.

(iii) Find matrix $\mathbf{M} = \mathbf{BA}$ and show that $\mathbf{MS}_0$ gives $\mathbf{S}_2$, the matrix which represents the order of the strands at Stage 2.

(iv) Find $\mathbf{M}^2$ and hence find the order of the strands at Stage 4.

(v) Calculate $\mathbf{M}^3$. What does this tell you?

3 Transformations

You are already familiar with several different types of transformation, including reflections, rotations and enlargements.

■ The original point, or shape, is called the **object**.

■ The new point, or shape, after the transformation, is called the **image**.

■ A transformation is a **mapping** of an object onto its image.

Some examples of transformations are illustrated in Figures 1.8 to 1.10 (note that the vertices of the image are denoted by the same letters with a dash, e.g. A′, B′).

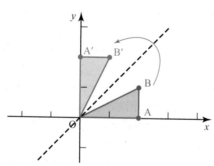

Figure 1.8 Reflection in the line $y = x$

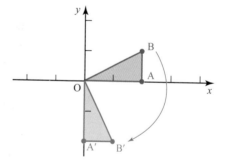

Figure 1.9 Rotation through 90° clockwise, centre O

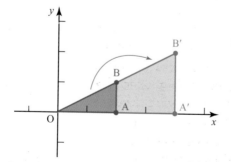

Figure 1.10 Enlargement centre O, scale factor 2

In this section, you will also meet the idea of

■ a **stretch** parallel to the x-axis or y-axis

and three-dimensional transformations where

■ a shape is reflected in the planes $x = 0$, $y = 0$ or $z = 0$
■ a shape is rotated about one of the three coordinate axes.

A transformation maps an object according to a rule and can be represented by a matrix (see next section). The effect of a transformation on an object can be found by looking at the effect it has on the **position vector** of the point $\begin{pmatrix} x \\ y \end{pmatrix}$,

i.e. the vector from the origin to the point (x, y). So, for example, to find the effect of a transformation on the point $(2, 3)$ you would look at the effect that the transformation matrix has on the position vector $\begin{pmatrix} 2 \\ 3 \end{pmatrix}$.

Vectors that have length or **magnitude** of 1 are called **unit vectors**.

In two dimensions, two unit vectors that are of particular interest are

$$\mathbf{i} = \begin{pmatrix} 1 \\ 0 \end{pmatrix}$$ – a unit vector in the direction of the x-axis

$$\mathbf{j} = \begin{pmatrix} 0 \\ 1 \end{pmatrix}$$ – a unit vector in the direction of the y-axis.

The equivalent unit vectors in three dimensions are

$$\mathbf{i} = \begin{pmatrix} 1 \\ 0 \\ 0 \end{pmatrix}$$ – a unit vector in the direction of the x-axis

$$\mathbf{j} = \begin{pmatrix} 0 \\ 1 \\ 0 \end{pmatrix}$$ – a unit vector in the direction of the y-axis

$$\mathbf{k} = \begin{pmatrix} 0 \\ 0 \\ 1 \end{pmatrix}$$ – a unit vector in the direction of the z-axis.

Finding the transformation represented by a given matrix

Start by looking at the effect of multiplying the unit vectors $\mathbf{i} = \begin{pmatrix} 1 \\ 0 \end{pmatrix}$

and $\mathbf{j} = \begin{pmatrix} 0 \\ 1 \end{pmatrix}$ by the matrix $\begin{pmatrix} -1 & 0 \\ 0 & -1 \end{pmatrix}$.

The image of $\begin{pmatrix} 1 \\ 0 \end{pmatrix}$ under this transformation is given by

$$\begin{pmatrix} -1 & 0 \\ 0 & -1 \end{pmatrix} \begin{pmatrix} 1 \\ 0 \end{pmatrix} = \begin{pmatrix} -1 \\ 0 \end{pmatrix}.$$

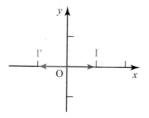

> **Note**
> The letter I is often used for the point (1, 0).

Figure 1.11

The image of $\begin{pmatrix} 0 \\ 1 \end{pmatrix}$ under the transformation is given by

$$\begin{pmatrix} -1 & 0 \\ 0 & -1 \end{pmatrix} \begin{pmatrix} 0 \\ 1 \end{pmatrix} = \begin{pmatrix} 0 \\ -1 \end{pmatrix}.$$

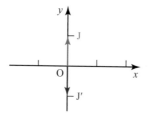

> **Note**
> The letter J is often used for the point (0, 1).

Figure 1.12

You can see from this that the matrix $\begin{pmatrix} -1 & 0 \\ 0 & -1 \end{pmatrix}$ represents a rotation, centre the origin, through 180°.

Example 1.4

Describe the transformations represented by the following matrices.

(i) $\begin{pmatrix} 0 & 1 \\ 1 & 0 \end{pmatrix}$ 　　　　(ii) $\begin{pmatrix} 2 & 0 \\ 0 & 2 \end{pmatrix}$

Solution

(i) $\begin{pmatrix} 0 & 1 \\ 1 & 0 \end{pmatrix} \begin{pmatrix} 1 \\ 0 \end{pmatrix} = \begin{pmatrix} 0 \\ 1 \end{pmatrix}$ 　　$\begin{pmatrix} 0 & 1 \\ 1 & 0 \end{pmatrix} \begin{pmatrix} 0 \\ 1 \end{pmatrix} = \begin{pmatrix} 1 \\ 0 \end{pmatrix}$

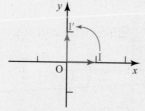 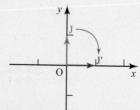

Figure 1.13　　　　**Figure 1.14**

The matrix $\begin{pmatrix} 0 & 1 \\ 1 & 0 \end{pmatrix}$ represents a reflection in the line $y = x$.

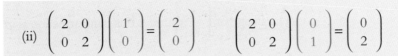

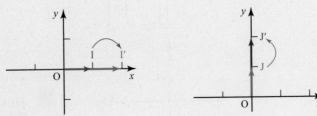

Figure 1.15   Figure 1.16

The matrix $\begin{pmatrix} 2 & 0 \\ 0 & 2 \end{pmatrix}$ represents an enlargement, centre the origin, scale factor 2.

You can see that the images of $\mathbf{i} = \begin{pmatrix} 1 \\ 0 \end{pmatrix}$ and $\mathbf{j} = \begin{pmatrix} 0 \\ 1 \end{pmatrix}$ are the two columns of the transformation matrix.

Finding the matrix that represents a given transformation

The connection between the images of the unit vectors i and j and the matrix representing the transformation provides a quick method for finding the matrix representing a transformation.

It is common to use the unit square with coordinates $O(0, 0)$, $I(1, 0)$, $P(1, 1)$ and $J(0, 1)$.

You can think about the images of the points I and J, and from this you can write down the images of the unit vectors $\mathbf{i}$ and $\mathbf{j}$.

This is done in the next example.

> **Hint**
>
> You may find it easier to see what the transformation is when you use a shape, like the unit square, rather than points or lines.

Example 1.5

By drawing a diagram to show the image of the unit square, find the matrices which represent each of the following transformations:

(i) a reflection in the x-axis

(ii) an enlargement of scale factor 3, centre the origin.

Solution

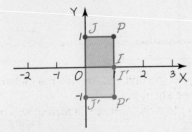

Figure 1.17

(i) You can see from Figure 1.17 that I $(1, 0)$ is mapped to itself and J $(0, 1)$ is mapped to J′ $(0, -1)$.

and the image of **J** is $\begin{pmatrix} 0 \\ -1 \end{pmatrix}$.

So the image of **I** is $\begin{pmatrix} 1 \\ 0 \end{pmatrix}$

So the matrix which represents a reflection in the x–axis is $\begin{pmatrix} 1 & 0 \\ 0 & -1 \end{pmatrix}$.

(ii)

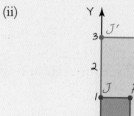

Figure 1.18

So the image of **I** is $\begin{pmatrix} 3 \\ 0 \end{pmatrix}$

You can see from Figure 1.18 that I $(1, 0)$ is mapped to I′ $(3, 0)$, and J $(0, 1)$ is mapped to J′ $(0, 3)$.

and the image of **J** is $\begin{pmatrix} 0 \\ 3 \end{pmatrix}$.

So the matrix which represents an enlargement, centre the origin, scale factor 3 is $\begin{pmatrix} 3 & 0 \\ 0 & 3 \end{pmatrix}$.

Discussion point

→ For a general transformation represented by the matrix $\begin{pmatrix} a & b \\ c & d \end{pmatrix}$, what are the images of the unit vectors $\begin{pmatrix} 1 \\ 0 \end{pmatrix}$ and $\begin{pmatrix} 0 \\ 1 \end{pmatrix}$?

→ What is the image of the origin $(0, 0)$?

ACTIVITY 1.3

Using the image of the unit square, find the matrix which represents a rotation of 45° anticlockwise about the origin.
Use your answer to write down the matrices which represent the following transformations:

(i) a rotation of 45° clockwise about the origin

(ii) a rotation of 135° anticlockwise about the origin.

Example 1.6

(i) Find the matrix which represents a rotation through angle θ anticlockwise about the origin.

(ii) Use your answer to find the matrix which represents a rotation of 60° anticlockwise about the origin.

Solution

(i) Figure 1.19 shows a rotation of angle θ anticlockwise about the origin.

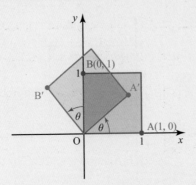

Figure 1.19

Call the coordinates of the point A' (p, q). Since the lines OA and OB are perpendicular, the coordinates of B' will be $(-q, p)$.

From the right-angled triangle with OA' as the hypotenuse, $\cos\theta = \frac{p}{1}$ and so $p = \cos\theta$.

Similarly $\sin\theta = \frac{q}{1}$ so $q = \sin\theta$.

So, the image point A' (p, q) has position vector $\begin{pmatrix} \cos\theta \\ \sin\theta \end{pmatrix}$ and the

image point B' $(-q, p)$ has position vector $\begin{pmatrix} -\sin\theta \\ \cos\theta \end{pmatrix}$.

Therefore, the matrix that represents a rotation of angle θ anticlockwise about the origin is $\begin{pmatrix} \cos\theta & -\sin\theta \\ \sin\theta & \cos\theta \end{pmatrix}$.

(ii) The matrix that represents an anticlockwise rotation of 60° about the

origin is $\begin{pmatrix} \cos 60° & -\sin 60° \\ \sin 60° & \cos 60° \end{pmatrix} = \begin{pmatrix} \dfrac{1}{2} & -\dfrac{\sqrt{3}}{2} \\ \dfrac{\sqrt{3}}{2} & \dfrac{1}{2} \end{pmatrix}$.

> **Discussion point**
> → What matrix would represent a rotation through angle θ clockwise about the origin?

TECHNOLOGY

You could use geometrical software to try different values of m and n.

ACTIVITY 1.4

Investigate the effect of the matrices:

(i) $\begin{pmatrix} 2 & 0 \\ 0 & 1 \end{pmatrix}$ (ii) $\begin{pmatrix} 1 & 0 \\ 0 & 5 \end{pmatrix}$

Describe the general transformation represented by the

matrices $\begin{pmatrix} m & 0 \\ 0 & 1 \end{pmatrix}$ and $\begin{pmatrix} 1 & 0 \\ 0 & n \end{pmatrix}$.

Activity 1.4 illustrates two important general results.

- The matrix $\begin{pmatrix} m & 0 \\ 0 & 1 \end{pmatrix}$ represents a stretch of scale factor m parallel to the x-axis.

- The matrix $\begin{pmatrix} 1 & 0 \\ 0 & n \end{pmatrix}$ represents a stretch of scale factor n parallel to the y-axis.

Summary of transformations in two dimensions

> **Note**
>
> All these transformations are examples of linear transformations. In a linear transformation, straight lines are mapped to straight lines, and the origin is mapped to itself.

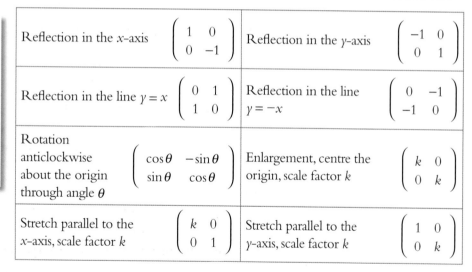

Reflection in the x-axis	$\begin{pmatrix} 1 & 0 \\ 0 & -1 \end{pmatrix}$	Reflection in the y-axis	$\begin{pmatrix} -1 & 0 \\ 0 & 1 \end{pmatrix}$
Reflection in the line $y = x$	$\begin{pmatrix} 0 & 1 \\ 1 & 0 \end{pmatrix}$	Reflection in the line $y = -x$	$\begin{pmatrix} 0 & -1 \\ -1 & 0 \end{pmatrix}$
Rotation anticlockwise about the origin through angle θ	$\begin{pmatrix} \cos\theta & -\sin\theta \\ \sin\theta & \cos\theta \end{pmatrix}$	Enlargement, centre the origin, scale factor k	$\begin{pmatrix} k & 0 \\ 0 & k \end{pmatrix}$
Stretch parallel to the x-axis, scale factor k	$\begin{pmatrix} k & 0 \\ 0 & 1 \end{pmatrix}$	Stretch parallel to the y-axis, scale factor k	$\begin{pmatrix} 1 & 0 \\ 0 & k \end{pmatrix}$

Transformations in three dimensions

When working with matrices, it is sometimes necessary to refer to a 'plane' – this is an infinite two-dimensional flat surface with no thickness. Figure 1.20 illustrates some common planes in three dimensions – the XY plane, the XZ plane and YZ plane. These three planes will be referred to when using matrices to represent some transformations in three dimensions. The plane XY can also be referred to as $z = 0$, since the z-coordinate would be zero for all points in the XY plane. Similarly, the XZ plane is referred to as $y = 0$ and the YZ plane as $x = 0$.

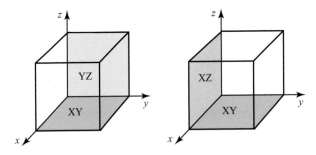

Figure 1.20

So far you have looked at transformations of sets of points from a plane (i.e. two dimensions) to the same plane. In a similar way, you can transform a set of points within three-dimensional space. You will look at reflections in the planes $x = 0$, $y = 0$ or $z = 0$, and rotations about one of the coordinate axes. Again, the matrix can be found algebraically or by considering the effect of the transformation on the three unit vectors

$$\mathbf{i} = \begin{pmatrix} 1 \\ 0 \\ 0 \end{pmatrix}, \mathbf{j} = \begin{pmatrix} 0 \\ 1 \\ 0 \end{pmatrix} \text{ and } \mathbf{k} = \begin{pmatrix} 0 \\ 0 \\ 1 \end{pmatrix}.$$

Think about reflecting an object in the plane $y = 0$. The plane $y = 0$ is the plane which contains the x- and z-axes. Figure 1.21 shows the effect of a reflection in the plane $y = 0$.

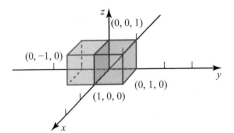

Figure 1.21

$$\mathbf{i} = \begin{pmatrix} 1 \\ 0 \\ 0 \end{pmatrix} \text{ maps to } \begin{pmatrix} 1 \\ 0 \\ 0 \end{pmatrix}, \mathbf{j} = \begin{pmatrix} 0 \\ 1 \\ 0 \end{pmatrix} \text{ maps to } \begin{pmatrix} 0 \\ -1 \\ 0 \end{pmatrix} \text{ and } \mathbf{k} = \begin{pmatrix} 0 \\ 0 \\ 1 \end{pmatrix} \text{ maps}$$

$$\text{to } \begin{pmatrix} 0 \\ 0 \\ 1 \end{pmatrix}.$$

The images of $\mathbf{i}$, $\mathbf{j}$ and $\mathbf{k}$ form the columns of the 3×3 transformation matrix.

$$\text{It is } \begin{pmatrix} 1 & 0 & 0 \\ 0 & -1 & 0 \\ 0 & 0 & 1 \end{pmatrix}.$$

| **Example 1.7** | Find the matrix that represents a rotation of 90° anticlockwise about the *x*-axis. |

Solution

A rotation of 90° anticlockwise about the *x*-axis is shown in Figure 1.22.

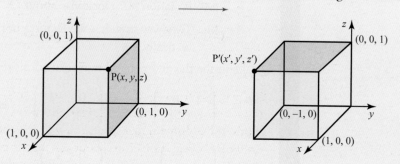

Note

Rotations are taken to be anticlockwise about the axis of rotation when looking along the axis from the positive end towards the origin.

Figure 1.22

Look at the effect of the transformation on the unit vectors **i**, **j** and **k**:

$$\mathbf{i} = \begin{pmatrix} 1 \\ 0 \\ 0 \end{pmatrix} \text{ maps to } \begin{pmatrix} 1 \\ 0 \\ 0 \end{pmatrix}, \mathbf{j} = \begin{pmatrix} 0 \\ 1 \\ 0 \end{pmatrix} \text{ maps to } \begin{pmatrix} 0 \\ 0 \\ 1 \end{pmatrix} \text{ and } \mathbf{k} = \begin{pmatrix} 0 \\ 0 \\ 1 \end{pmatrix}$$

$$\text{maps to } \begin{pmatrix} 0 \\ -1 \\ 0 \end{pmatrix}.$$

The images of **i**, **j** and **k** form the columns of the 3 × 3 transformation matrix.

The matrix is $\begin{pmatrix} 1 & 0 & 0 \\ 0 & 0 & -1 \\ 0 & 1 & 0 \end{pmatrix}$.

| **Exercise 1.3** | |

① Figure 1.23 shows a triangle with vertices at O, A(1, 2) and B(0, 2).

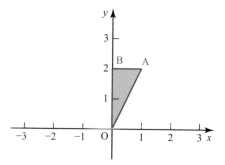

Figure 1.23

For each of the transformations below

(a) draw a diagram to show the effect of the transformation on triangle OAB

(b) give the coordinates of A′ and B′, the images of points A and B

(c) find expressions for *x*′ and *y*′, the coordinates of P′, the image of a general point P(*x*, *y*)

(d) find the matrix which represents the transformation.

(i) Enlargement, centre the origin, scale factor 3

(ii) Reflection in the x-axis

(iii) Reflection in the line $x + y = 0$

(iv) Rotation 90° clockwise about O

(v) Two-way stretch, scale factor 3 horizontally and scale factor $\frac{1}{2}$ vertically.

② Describe the geometrical transformations represented by these matrices.

(i) $\begin{pmatrix} 1 & 0 \\ 0 & -1 \end{pmatrix}$ (ii) $\begin{pmatrix} 0 & -1 \\ -1 & 0 \end{pmatrix}$ (iii) $\begin{pmatrix} 2 & 0 \\ 0 & 3 \end{pmatrix}$

(iv) $\begin{pmatrix} 4 & 0 \\ 0 & 4 \end{pmatrix}$ (v) $\begin{pmatrix} 0 & 1 \\ -1 & 0 \end{pmatrix}$

③ Each of the following matrices represents a rotation about the origin. Find the angle and direction of rotation in each case.

(i) $\begin{pmatrix} \dfrac{1}{2} & -\dfrac{\sqrt{3}}{2} \\ \dfrac{\sqrt{3}}{2} & \dfrac{1}{2} \end{pmatrix}$ (ii) $\begin{pmatrix} 0.574 & -0.819 \\ 0.819 & 0.574 \end{pmatrix}$

(iii) $\begin{pmatrix} -\dfrac{1}{\sqrt{2}} & \dfrac{1}{\sqrt{2}} \\ -\dfrac{1}{\sqrt{2}} & -\dfrac{1}{\sqrt{2}} \end{pmatrix}$ (iv) $\begin{pmatrix} -\dfrac{\sqrt{3}}{2} & -\dfrac{1}{2} \\ \dfrac{1}{2} & -\dfrac{\sqrt{3}}{2} \end{pmatrix}$

④ Find the matrix that represents each of the following transformations in three dimensions.

(i) Rotation of 90° anticlockwise about the z-axis

(ii) Reflection in the plane $y = 0$

(iii) Rotation of 180° about the x-axis

(iv) Rotation of 270° anticlockwise about the y-axis

⑤ The unit square OABC has its vertices at $(0, 0)$, $(1, 0)$, $(1, 1)$ and $(0, 1)$.

OABC is mapped to OA′B′C′ by the transformation defined by the matrix $\begin{pmatrix} 4 & 3 \\ 5 & 4 \end{pmatrix}$.

Find the coordinates of A′, B′ and C′ and show that the area of the shape has not been changed by the transformation.

⑥ The transformation represented by the matrix $\mathbf{M} = \begin{pmatrix} 1 & 2 \\ 0 & 1 \end{pmatrix}$ is applied to

the triangle ABC with vertices A$(-1, 1)$, B$(1, -1)$ and C$(-1, -1)$.

(i) Draw a diagram showing the triangle ABC and its image A′B′C′.

(ii) Find the gradient of the line A′C′ and explain how this relates to the matrix $\mathbf{M}$.

⑦ Describe the transformations represented by these matrices.

(i) $\begin{pmatrix} 1 & 0 & 0 \\ 0 & 0 & 1 \\ 0 & -1 & 0 \end{pmatrix}$ (ii) $\begin{pmatrix} 3 & 0 & 0 \\ 0 & 3 & 0 \\ 0 & 0 & 3 \end{pmatrix}$ (iii) $\begin{pmatrix} 1 & 0 & 0 \\ 0 & 1 & 0 \\ 0 & 0 & -1 \end{pmatrix}$ (iv) $\begin{pmatrix} 2 & 0 & 0 \\ 0 & 3 & 0 \\ 0 & 0 & \frac{1}{2} \end{pmatrix}$

⑧ Find the matrices that would represent

(i) a reflection in the plane $z = 0$

(ii) a rotation of $180°$ about the y-axis.

⑨ A transformation maps P to P' as follows:

■ Each point is mapped on to the line $y = x$.

■ The line joining a point to its image is parallel to the y-axis.

Find the coordinates of the image of the point (x, y) and hence show that this transformation can be represented by means of a matrix.

What is that matrix?

⑩ A square has corners with coordinates A$(1, 0)$, B$(1, 1)$, C$(0, 1)$ and O$(0, 0)$. It is to be transformed into another quadrilateral in the first quadrant of the coordinate grid.

Find a matrix which would transform the square into

(i) a rectangle with one vertex at the origin, the sides lie along the axes and one side of length is 5 units

(ii) a rhombus with one vertex at the origin, two angles of $45°$ and side lengths of $\sqrt{2}$ units; one of the sides lies along an axis

(iii) a parallelogram with one vertex at the origin and two angles of $30°$; one of the longest sides lies along an axis and has length 7 units; the shortest sides have length 3 units.

Is there more than one possibility for any of these matrices? If so, write down alternative matrices that satisfy the same description.

4 Successive transformations

Figure 1.24 shows the effect of two successive transformations on a triangle. The transformation A represents a reflection in the x-axis. A maps the point P to the point A(P).

The transformation B represents a rotation of $90°$ anticlockwise about O. When you apply B to the image formed by A, the point A(P) is mapped to the point B(A(P)). This is abbreviated to BA(P).

Note

Notice that a transformation written as BA means 'carry out A, then carry out B'.

This process is sometimes called **composition of transformations**.

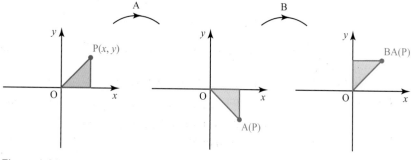

Figure 1.24

Note

A transformation is often denoted by a capital letter. The matrix representing this transformation is usually denoted by the same letter, in bold upright.

Discussion point

➜ Look at Figure 1.24 and compare the original triangle with the final image after both transformations.

(i) Describe the single transformation represented by BA.

(ii) Write down the matrices which represent the transformations A and B. Calculate the matrix product **BA** and comment on your answer.

In general, the matrix for a composite transformation is found by multiplying the matrices of the individual transformations in reverse order. So, for two transformations the matrix representing the first transformation is on the right and the matrix for the second transformation is on the left. For n transformations $T_1, T_2, \ldots, T_{n-1}, T_n$, the matrix product would be $\mathbf{T}_n \mathbf{T}_{n-1} \cdots \mathbf{T}_2 \mathbf{T}_1$.

You will prove this result for two transformations in Activity 1.5.

TECHNOLOGY

If you have access to geometrical software, you could investigate this using several different matrices for **T** and **S**.

ACTIVITY 1.5

The transformations T and S are represented by the matrices $\mathbf{T} = \begin{pmatrix} a & c \\ b & d \end{pmatrix}$ and $\mathbf{S} = \begin{pmatrix} p & r \\ q & s \end{pmatrix}$.

T is applied to the point P with position vector $\mathbf{p} = \begin{pmatrix} x \\ y \end{pmatrix}$. The image of P is P′.

S is then applied to the point P′. The image of P′ is P″. This is illustrated in Figure 1.25.

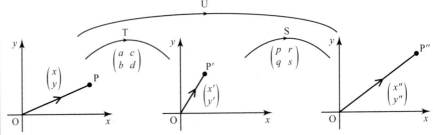

Figure 1.25

(i) Find the position vector $\begin{pmatrix} x' \\ y' \end{pmatrix}$ of P′ by calculating the matrix product $\mathbf{T} \begin{pmatrix} x \\ y \end{pmatrix}$.

(ii) Find the position vector $\begin{pmatrix} x'' \\ y'' \end{pmatrix}$ of P″ by calculating the matrix product $\mathbf{S} \begin{pmatrix} x' \\ y' \end{pmatrix}$.

(iii) Find the matrix product $\mathbf{U} = \mathbf{ST}$ and show that $\mathbf{U} \begin{pmatrix} x \\ y \end{pmatrix}$ is the same as $\begin{pmatrix} x'' \\ y'' \end{pmatrix}$.

Discussion point

➜ How can you use the idea of successive transformations to explain the associativity of matrix multiplication $(\mathbf{AB})\mathbf{C} = \mathbf{A}(\mathbf{BC})$?

Proving results in trigonometry

If you carry out a rotation about the origin through angle θ, followed by a rotation about the origin through angle ϕ, then this is equivalent to a single rotation about the origin through angle $\theta + \phi$. Rotations are measured in an anticlockwise direction unless you are told otherwise. Using matrices to represent these transformations allows you to prove the formulae for $\sin(\theta + \phi)$ and $\cos(\theta + \phi)$ given on page 201. This is done in Activity 1.6.

ACTIVITY 1.6

(i) Write down the matrix $\mathbf{A}$ representing a rotation about the origin through angle θ, and the matrix $\mathbf{B}$ representing a rotation about the origin through angle ϕ.

(ii) Find the matrix $\mathbf{BA}$, representing a rotation about the origin through angle θ, followed by a rotation about the origin through angle ϕ.

(iii) Write down the matrix $\mathbf{C}$ representing a rotation about the origin through angle $\theta + \phi$.

(iv) By equating $\mathbf{C}$ to $\mathbf{BA}$, write down expressions for $\sin(\theta + \phi)$ and $\cos(\theta + \phi)$.

(v) Explain why $\mathbf{BA} = \mathbf{AB}$ in this case.

Example 1.8

(i) Write down the matrix $\mathbf{A}$ which represents a rotation of $135°$ about the origin.

(ii) Write down the matrices $\mathbf{B}$ and $\mathbf{C}$ which represent rotations of $45°$ and $90°$ respectively about the origin. Find the matrix $\mathbf{BC}$ and verify that $\mathbf{A} = \mathbf{BC}$.

(iii) Calculate the matrix $\mathbf{B}^3$ and comment on your answer.

Solution

(i) $\mathbf{A} = \begin{pmatrix} -\dfrac{1}{\sqrt{2}} & -\dfrac{1}{\sqrt{2}} \\ \dfrac{1}{\sqrt{2}} & -\dfrac{1}{\sqrt{2}} \end{pmatrix}$

(ii) $\mathbf{B} = \begin{pmatrix} \dfrac{1}{\sqrt{2}} & -\dfrac{1}{\sqrt{2}} \\ \dfrac{1}{\sqrt{2}} & \dfrac{1}{\sqrt{2}} \end{pmatrix}, \mathbf{C} = \begin{pmatrix} 0 & -1 \\ 1 & 0 \end{pmatrix}$

$\mathbf{BC} = \begin{pmatrix} \dfrac{1}{\sqrt{2}} & -\dfrac{1}{\sqrt{2}} \\ \dfrac{1}{\sqrt{2}} & \dfrac{1}{\sqrt{2}} \end{pmatrix}\begin{pmatrix} 0 & -1 \\ 1 & 0 \end{pmatrix} = \begin{pmatrix} -\dfrac{1}{\sqrt{2}} & -\dfrac{1}{\sqrt{2}} \\ \dfrac{1}{\sqrt{2}} & -\dfrac{1}{\sqrt{2}} \end{pmatrix} = \mathbf{A}$

(iii) $\mathbf{B}^3 = \begin{pmatrix} \dfrac{1}{\sqrt{2}} & -\dfrac{1}{\sqrt{2}} \\ \dfrac{1}{\sqrt{2}} & \dfrac{1}{\sqrt{2}} \end{pmatrix}\begin{pmatrix} \dfrac{1}{\sqrt{2}} & -\dfrac{1}{\sqrt{2}} \\ \dfrac{1}{\sqrt{2}} & \dfrac{1}{\sqrt{2}} \end{pmatrix}\begin{pmatrix} \dfrac{1}{\sqrt{2}} & -\dfrac{1}{\sqrt{2}} \\ \dfrac{1}{\sqrt{2}} & \dfrac{1}{\sqrt{2}} \end{pmatrix} = \begin{pmatrix} -\dfrac{1}{\sqrt{2}} & -\dfrac{1}{\sqrt{2}} \\ \dfrac{1}{\sqrt{2}} & -\dfrac{1}{\sqrt{2}} \end{pmatrix}$

This verifies that three successive anticlockwise rotations of $45°$ about the origin is equivalent to a single anticlockwise rotation of $135°$ about the origin.

1

Chapter 1 Matrices and transformations

25

① $\mathbf{A} = \begin{pmatrix} 3 & 0 \\ 0 & 3 \end{pmatrix}$, $\mathbf{B} = \begin{pmatrix} 0 & -1 \\ 1 & 0 \end{pmatrix}$, $\mathbf{C} = \begin{pmatrix} 1 & 0 \\ 0 & -1 \end{pmatrix}$ and $\mathbf{D} = \begin{pmatrix} 0 & 1 \\ 1 & 0 \end{pmatrix}$.

(i) Describe the transformations that are represented by matrices $\mathbf{A}, \mathbf{B}, \mathbf{C}$ and $\mathbf{D}$.

(ii) Find the following matrix products and describe the single transformation represented in each case:

(a) $\mathbf{BC}$ (b) $\mathbf{CB}$ (c) $\mathbf{DC}$ (d) $\mathbf{A}^2$ (e) $\mathbf{BCB}$ (f) $\mathbf{DC}^2\mathbf{D}$

(iii) Write down two other matrix products, using the matrices $\mathbf{A}, \mathbf{B}, \mathbf{C}$ and $\mathbf{D}$, which would produce the same single transformation as $\mathbf{DC}^2\mathbf{D}$.

② The matrix $\mathbf{X}$ represents a reflection in the x-axis.

The matrix $\mathbf{Y}$ represents a reflection in the y-axis.

(i) Write down the matrices $\mathbf{X}$ and $\mathbf{Y}$.

(ii) Find the matrix $\mathbf{XY}$ and describe the transformation it represents.

(iii) Find the matrix $\mathbf{YX}$.

(iv) Explain geometrically why $\mathbf{XY} = \mathbf{YX}$ in this case.

③ The matrix $\mathbf{P}$ represents a rotation of 180° about the origin.

The matrix $\mathbf{Q}$ represents a reflection in the line $y = x$.

(i) Write down the matrices $\mathbf{P}$ and $\mathbf{Q}$.

(ii) Find the matrix $\mathbf{PQ}$ and describe the transformation it represents.

(iii) Find the matrix $\mathbf{QP}$.

(iv) Explain geometrically why $\mathbf{PQ} = \mathbf{QP}$ in this case.

④ In three dimensions, the four matrices $\mathbf{J, K, L}$ and $\mathbf{M}$ represent transformations as follows:

$\mathbf{J}$ represents a reflection in the plane $z = 0$.

$\mathbf{K}$ represents a rotation of 90° about the x-axis.

$\mathbf{L}$ represents a reflection in the plane $x = 0$.

$\mathbf{M}$ represents a rotation of 90° about the y-axis.

(i) Write down the matrices $\mathbf{J, K, L}$ and $\mathbf{M}$.

(ii) Write down matrix products which would represent the single transformations obtained by each of the following combinations of transformations.

(a) A reflection in the plane $z = 0$ followed by a reflection in the plane $x = 0$

(b) A reflection in the plane $z = 0$ followed by a rotation of 90° about the y-axis

(c) A rotation of 90° about the x-axis followed by a second rotation of 90° about the x-axis

(d) A rotation of 90° about the x-axis followed by a reflection in the plane $x = 0$ followed by a reflection in the plane $z = 0$

⑤ The transformations R and S are represented by the matrices

$$\mathbf{R} = \begin{pmatrix} 2 & -1 \\ 1 & 3 \end{pmatrix} \text{ and } \mathbf{S} = \begin{pmatrix} 3 & 0 \\ -2 & 4 \end{pmatrix}.$$

(i)   Find the matrix which represents the transformation RS.

(ii) Find the image of the point $(3, -2)$ under the transformation RS.

⑥ The transformation represented by $C = \begin{pmatrix} 0 & 3 \\ -1 & 0 \end{pmatrix}$ is equivalent to a single

transformation B followed by a single transformation A. Give geometrical descriptions of a pair of possible transformations B and A and state the matrices that represent them.
Comment on the order in which the transformations are performed.

⑦ Find the matrix $\mathbf{X}$ which represents a rotation of 135° about the origin followed by a reflection in the y-axis.

Explain why matrix $\mathbf{X}$ cannot represent a rotation about the origin.

⑧ Find the matrix $\mathbf{Y}$ which represents a reflection in the plane $y = 0$ followed by a rotation of 90° about the z-axis.

⑨ (i)   Write down the matrix $\mathbf{P}$ which represents a stretch of scale factor 2 parallel to the y-axis.

(ii) The matrix $\mathbf{Q} = \begin{pmatrix} 5 & 0 \\ 0 & -1 \end{pmatrix}$. Write down the two single

transformations which are represented by the matrix $\mathbf{Q}$.

(iii) Find the matrix $\mathbf{PQ}$. Write a list of the three transformations which are represented by the matrix $\mathbf{PQ}$. In how many different orders could the three transformations occur?

(iv) Find the matrix $\mathbf{R}$ for which the matrix product $\mathbf{RPQ}$ would transform an object to its original position.

⑩ There are two basic types of four-terminal electrical networks, as shown in Figure 1.26.

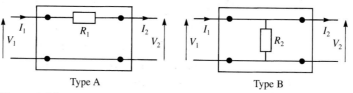

Figure 1.26

In Type A the output voltage V_2 and current I_2 are related to the input voltage V_1 and current I_1 by the simultaneous equations:

$$V_2 = V_1 - I_1 R_1$$
$$I_2 = I_1$$

The simultaneous equations can be written as $\begin{pmatrix} V_2 \\ I_2 \end{pmatrix} = \mathbf{A} \begin{pmatrix} V_1 \\ I_1 \end{pmatrix}$.

(i)   Find the matrix $\mathbf{A}$.

In Type B the corresponding simultaneous equations are:

$$V_2 = V_1$$
$$I_2 = I_1 - \frac{V_1}{R_2}$$

(ii) Write down the matrix **B** which represents the effect of a Type B network.

(iii) Find the matrix which represents the effect of Type A followed by Type B.

(iv) Is the effect of Type B followed by Type A the same as the effect of Type A followed by Type B?

⑪ The matrix **B** represents a rotation of 45° anticlockwise about the origin.

$$\mathbf{B} = \begin{pmatrix} \dfrac{1}{\sqrt{2}} & -\dfrac{1}{\sqrt{2}} \\ \dfrac{1}{\sqrt{2}} & \dfrac{1}{\sqrt{2}} \end{pmatrix}, \mathbf{D} = \begin{pmatrix} a & -b \\ b & a \end{pmatrix} \text{ where } a \text{ and } b \text{ are positive real numbers}$$

Given that $\mathbf{D}^2 = \mathbf{B}$, find exact values for a and b. Write down the transformation represented by the matrix **D**. What do the exact values a and b represent?

5 Invariance

Invariant points

> **Discussion points**
> → In a reflection, are there any points which map to themselves?
> → In a rotation, are there any points which map to themselves?

Points which map to themselves under a transformation are called **invariant points**. The origin is always an invariant point under a transformation that can be represented by a matrix, as the following statement is always true:

$$\begin{pmatrix} a & b \\ c & d \end{pmatrix} \begin{pmatrix} 0 \\ 0 \end{pmatrix} = \begin{pmatrix} 0 \\ 0 \end{pmatrix}$$

More generally, a point (x, y) is invariant if it satisfies the matrix equation:

$$\begin{pmatrix} a & b \\ c & d \end{pmatrix} \begin{pmatrix} x \\ y \end{pmatrix} = \begin{pmatrix} x \\ y \end{pmatrix}$$

For example, the point $(-2, 2)$ is invariant under the transformation represented

by the matrix $\begin{pmatrix} 6 & 5 \\ 2 & 3 \end{pmatrix}$: $\begin{pmatrix} 6 & 5 \\ 2 & 3 \end{pmatrix} \begin{pmatrix} -2 \\ 2 \end{pmatrix} = \begin{pmatrix} -2 \\ 2 \end{pmatrix}$

Example 1.9

M is the matrix $\begin{pmatrix} 2 & -1 \\ 1 & 0 \end{pmatrix}$.

(i) Show that $(5, 5)$ is an invariant point under the transformation represented by **M**.

(ii) What can you say about the invariant points under this transformation?

Solution

(i) $\begin{pmatrix} 2 & -1 \\ 1 & 0 \end{pmatrix}\begin{pmatrix} 5 \\ 5 \end{pmatrix} = \begin{pmatrix} 5 \\ 5 \end{pmatrix}$ so $(5, 5)$ is an invariant point under the

transformation represented by **M**.

(ii) Suppose the point $\begin{pmatrix} x \\ y \end{pmatrix}$ maps to itself. Then

$$\begin{pmatrix} 2 & -1 \\ 1 & 0 \end{pmatrix}\begin{pmatrix} x \\ y \end{pmatrix} = \begin{pmatrix} x \\ y \end{pmatrix}$$

$$\begin{pmatrix} 2x - y \\ x \end{pmatrix} = \begin{pmatrix} x \\ y \end{pmatrix}$$

> Both equations simplify to $y = x$.

$\Leftrightarrow 2x - y = x$ and $x = y$.

So the invariant points of the transformation are all the points on the line $y = x$.

> These points all have the form (λ, λ). The point $(5, 5)$ is just one of the points on this line.

The simultaneous equations in Example 1.9 were equivalent and so all the invariant points were on a straight line. Generally, any matrix equation set up to find the invariant points will lead to two equations of the form $ax + by = 0$, which can also be expressed in the form $y = -\dfrac{ax}{b}$. These equations may be equivalent, in which case this is a line of invariant points. If the two equations are not equivalent, the origin is the only point which satisfies both equations, and so this is the only invariant point.

Invariant lines

A line AB is known as an **invariant line** under a transformation if the image of every point on AB is also on AB. It is important to note that it is not necessary for each of the points to map to itself; it can map to itself or to some other point on the line AB.

Sometimes it is easy to spot which lines are invariant. For example, in Figure 1.27 the position of the points A–F and their images A′–F′ show that the transformation is a reflection in the line *l*. So every point on *l* maps onto itself and *l* is a **line of invariant points**.

Look at the lines perpendicular to the mirror line in Figure 1.27, for example the line ABB′A′. Any point on one of these lines maps onto another point on the same line. Such a line is invariant but it is not a line of invariant points.

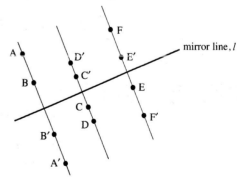

Figure 1.27

Example 1.10

Find the invariant lines of the transformation given by the matrix $\mathbf{M} = \begin{pmatrix} 5 & 1 \\ 2 & 4 \end{pmatrix}$.

Solution

Suppose the invariant line has the form $y = mx + c$

> Let the original point be (x, y) and the image point be (x', y').

$$\begin{pmatrix} x' \\ y' \end{pmatrix} = \begin{pmatrix} 5 & 1 \\ 2 & 4 \end{pmatrix} \begin{pmatrix} x \\ y \end{pmatrix} \Leftrightarrow x' = 5x + y \text{ and } y' = 2x + 4y$$

$$\Leftrightarrow \begin{cases} x' = 5x + mx + c = (5 + m)x + c \\ y' = 2x + 4(mx + c) = (2 + 4m)x + 4c \end{cases}$$

> Using $y = mx + c$.

As the line is invariant, (x', y') also lies on the line, so $y' = mx' + c$.

Therefore,

$$(2 + 4m)x + 4c = m[(5 + m)x + c] + c$$

$$\Leftrightarrow 0 = (m^2 + m - 2)x + (m - 3)c$$

For the left-hand side to equal zero, both $m^2 + m - 2 = 0$ and $(m - 3)c = 0$.

$$(m - 1)(m + 2) = 0 \Leftrightarrow m = 1 \text{ or } m = -2$$

and

$$(m - 3)c = 0 \Leftrightarrow m = 3 \text{ or } c = 0$$

> $m = 3$ is not a viable solution as $m^2 + m - 2 \neq 0$.

So, there are two possible solutions for the invariant line:

$$m = 1, c = 0 \Leftrightarrow y = x$$

or

$$m = -2, c = 0 \Leftrightarrow y = -2x$$

Figure 1.28 shows the effect of this transformation, together with its invariant lines.

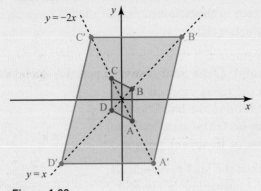

Figure 1.28

① Find the invariant points under the transformations represented by the following matrices.

(i) $\begin{pmatrix} -1 & -1 \\ 2 & 2 \end{pmatrix}$ (ii) $\begin{pmatrix} 3 & 4 \\ 1 & 2 \end{pmatrix}$ (iii) $\begin{pmatrix} 4 & 1 \\ 6 & 3 \end{pmatrix}$ (iv) $\begin{pmatrix} 7 & -4 \\ 3 & -1 \end{pmatrix}$

② What lines, if any, are invariant under the following transformations?

(i)   Enlargement, centre the origin
(ii) Rotation through 180° about the origin
(iii) Rotation through 90° about the origin
(iv) Reflection in the line $y = x$
(v)   Reflection in the line $y = -x$

③ Figure 1.29 shows the effect on the unit square of a transformation represented by $\mathbf{A} = \begin{pmatrix} 0.6 & 0.8 \\ 0.8 & -0.6 \end{pmatrix}$.

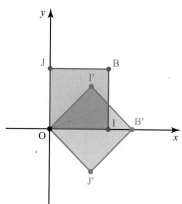

Figure 1.29

(i)   Find three points which are invariant under this transformation.
(ii) Given that this transformation is a reflection, write down the equation of the mirror line.
(iii) Using your answer to part (ii), write down the equation of an invariant line, other than the mirror line, under this reflection.
(iv) Justify your answer to part (iii) algebraically.

④ For the matrix $\mathbf{M} = \begin{pmatrix} 4 & 11 \\ 11 & 4 \end{pmatrix}$

(i)   show that the origin is the only invariant point
(ii) find the invariant lines of the transformation represented by $\mathbf{M}$.

⑤ (i)   Find the invariant lines of the transformation given by the matrix $\begin{pmatrix} 3 & 4 \\ 9 & -2 \end{pmatrix}$.

(ii) Draw a diagram to show the effect of the transformation on the unit square, and show the invariant lines on your diagram.

⑥ For the matrix $\mathbf{M} = \begin{pmatrix} 0 & 1 \\ -1 & 2 \end{pmatrix}$

   (i)   find the line of invariant points of the transformation given by $\mathbf{M}$

   (ii) find the invariant lines of the transformation

   (iii) draw a diagram to show the effect of the transformation on the unit square.

⑦ The matrix $\begin{pmatrix} \dfrac{1-m^2}{1+m^2} & \dfrac{2m}{1+m^2} \\ \dfrac{2m}{1+m^2} & \dfrac{m^2-1}{1+m^2} \end{pmatrix}$ represents a reflection in the line $y = mx$.

Prove that the line $y = mx$ is a line of invariant points.

⑧ The transformation T maps $\begin{pmatrix} x \\ y \end{pmatrix}$ to $\begin{pmatrix} a & b \\ c & d \end{pmatrix}\begin{pmatrix} x \\ y \end{pmatrix}$.

Show that invariant points other than the origin exist if $ad - bc = a + d - 1$.

⑨ T is a translation of the plane by the vector $\begin{pmatrix} a \\ b \end{pmatrix}$. The point (x, y) is mapped to the point (x', y').

   (i)   Write down equations for x' and y' in terms of x and y.

   (ii) Verify that $\begin{pmatrix} x' \\ y' \\ z' \end{pmatrix} = \begin{pmatrix} 1 & 0 & a \\ 0 & 1 & b \\ 0 & 0 & 1 \end{pmatrix}\begin{pmatrix} x \\ y \\ 1 \end{pmatrix}$ produces the same equations as those obtained in part (i).

The point (X, Y) is the image of the point (x, y) under the combined transformation TM where

$$\begin{pmatrix} X \\ Y \\ 1 \end{pmatrix} = \begin{pmatrix} -0.6 & 0.8 & a \\ 0.8 & 0.6 & b \\ 0 & 0 & 1 \end{pmatrix}\begin{pmatrix} x \\ y \\ 1 \end{pmatrix}$$

   (iii) (a) Show that if $a = -4$ and $b = 2$ then $(0, 5)$ is an invariant point of TM.

          (b) Show that if $a = 2$ and $b = 1$ then TM has no invariant point.

          (c) Find a relationship between a and b that must be satisfied if TM is to have any invariant points.

LEARNING OUTCOMES

When you have completed this chapter you should be able to:

➤ understand what is meant by the terms order of a matrix, square matrix, identity matrix, zero matrix and equal matrices

➤ add and subtract matrices of the same order

➤ multiply a matrix by a scalar

➤ know when two matrices are conformable for multiplication, and be able to multiply conformable matrices

➤ use a calculator to carry out matrix calculations

➤ know that matrix multiplication is associative but not commutative

➤ find the matrix associated with a linear transformation in two dimensions:

  ➤ reflections in the coordinate axes and the lines $y = \pm x$

  ➤ rotations about the origin

  ➤ enlargements centre the origin

  ➤ stretches parallel to the coordinate axes

➤ find the matrix associated with a linear transformation in three dimensions:

  ➤ reflection in $x = 0$, $y = 0$ or $z = 0$

  ➤ rotations through multiples of 90° about the x, y or z axes

➤ understand successive transformations in two dimensions and the connection with matrix multiplication

➤ find the invariant points for a linear transformation

➤ find the invariant lines for a linear transformation.

KEY POINTS

1 A matrix is a rectangular array of numbers or letters.

2 The shape of a matrix is described by its order. A matrix with r rows and c columns has order $r \times c$.

3 A matrix with the same number of rows and columns is called a square matrix.

4 The matrix $\mathbf{O} = \begin{pmatrix} 0 & 0 \\ 0 & 0 \end{pmatrix}$ is known as the 2×2 zero matrix. Zero matrices can be of any order.

5 A matrix of the form $\mathbf{I} = \begin{pmatrix} 1 & 0 \\ 0 & 1 \end{pmatrix}$ is known as an identity matrix. All identity matrices are square, with 1s on the leading diagonal and zeros elsewhere.

6 Matrices can be added or subtracted if they have the same order.

7 Two matrices $\mathbf{A}$ and $\mathbf{B}$ can be multiplied to give matrix $\mathbf{AB}$ if their orders are of the form $p \times q$ and $q \times r$ respectively. The resulting matrix will have the order $p \times r$.

8 Matrix multiplication:

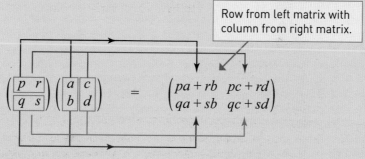

Row from left matrix with column from right matrix.

$$\left(\begin{array}{cc} p & r \\ \hline q & s \end{array}\right)\left(\begin{array}{c|c} a & c \\ b & d \end{array}\right) = \begin{pmatrix} pa + rb & pc + rd \\ qa + sb & qc + sd \end{pmatrix}$$

Figure 1.30

9 Matrix addition and multiplication are associative:
$$\mathbf{A} + (\mathbf{B} + \mathbf{C}) = (\mathbf{A} + \mathbf{B}) + \mathbf{C}$$
$$\mathbf{A}(\mathbf{BC}) = (\mathbf{AB})\mathbf{C}$$

10 Matrix addition is commutative but matrix multiplication is generally not commutative:

$$\mathbf{A} + \mathbf{B} = \mathbf{B} + \mathbf{A}$$

$$\mathbf{AB} \neq \mathbf{BA}$$

11 The matrix $\mathbf{M} = \begin{pmatrix} a & b \\ c & d \end{pmatrix}$ represents the transformation which maps the point with position vector $\begin{pmatrix} x \\ y \end{pmatrix}$ to the point with position vector $\begin{pmatrix} ax + by \\ cx + dy \end{pmatrix}$.

12 A list of the matrices representing common transformations, including rotations, reflections, enlargements and stretches, is given on page 19.

13 Under the transformation represented by $\mathbf{M}$, the image of $\mathbf{i} = \begin{pmatrix} 1 \\ 0 \end{pmatrix}$ is the first column of $\mathbf{M}$ and the image of $\mathbf{j} = \begin{pmatrix} 0 \\ 1 \end{pmatrix}$ is the second column of $\mathbf{M}$.

Similarly, in three dimensions the images of the unit vectors $\mathbf{i} = \begin{pmatrix} 1 \\ 0 \\ 0 \end{pmatrix}$, $\mathbf{j} = \begin{pmatrix} 0 \\ 1 \\ 0 \end{pmatrix}$ and $\mathbf{k} = \begin{pmatrix} 0 \\ 0 \\ 1 \end{pmatrix}$ are the first, second and third columns of the transformation matrix.

14 The composite of the transformation represented by M followed by that represented by N is represented by the matrix product $\mathbf{NM}$.

15 If (x, y) is an invariant point under a transformation represented by the matrix $\mathbf{M}$, then $\mathbf{M} \begin{pmatrix} x \\ y \end{pmatrix} = \begin{pmatrix} x \\ y \end{pmatrix}$.

16 A line AB is known as an invariant line under a transformation if the image of every point on AB is also on AB.

FUTURE USES

■ Work on matrices is developed further in Chapter 7 'Matrices and their inverses'.

Introduction to complex numbers

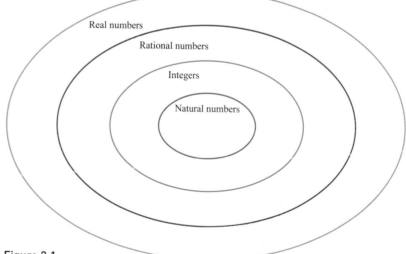

Figure 2.1

Discussion points

→ What is the meaning of each of the terms shown in Figure 2.1?

→ Suggest two numbers that could be placed in each part of the diagram.

1 Extending the number system

The number system we use today has taken thousands of years to develop. To classify the different types of numbers used in mathematics the following letter symbols are used:

Discussion point

➜ Why is there no set shown on the diagram for irrational numbers?

$\mathbb{N}$	Natural numbers
$\mathbb{Z}$	Integers
$\mathbb{Q}$	Rational numbers
$\bar{\mathbb{Q}}$	Irrational numbers
$\mathbb{R}$	Real numbers

You may have noticed that some of these sets of numbers fit within the other sets. This can be seen in Figure 2.1.

ACTIVITY 2.1

On a copy of Figure 2.1 write the following numbers in the correct positions.

$7 \quad \sqrt{5} \quad -13 \quad \dfrac{227}{109} \quad -\sqrt{5} \quad 3.1415 \quad \pi \quad 0.33 \quad 0.\dot{3}$

What are complex numbers?

ACTIVITY 2.2

Solve each of these equations and decide which set of numbers the roots belong to in each case.

(i) $x + 7 = 9$ (ii) $7x = 9$ (iii) $x^2 = 9$

(iv) $x + 10 = 9$ (v) $x^2 + 7x = 0$

Now think about the equation $x^2 + 9 = 0.$

You could rewrite it as $x^2 = -9$. However, since the square of every real number is positive or zero, there is no real number with a square of -9. This is an example of a quadratic equation which, up to now, you would have classified as having 'no real roots'.

> Writing this quadratic equation as $x^2 + 0x + 9 = 0$ and calculating the discriminant for this quadratic gives $b^2 - 4ac = -36$ which is less than zero.

Prior knowledge

You should know how to solve quadratic equations using the quadratic formula.

The existence of such equations was recognised for hundreds of years, in the same way that Greek mathematicians had accepted that $x + 10 = 9$ had no solution; the concept of a negative number had yet to be developed. The number system has expanded as mathematicians increased the range of mathematical problems they wanted to tackle.

You can solve the equation $x^2 + 9 = 0$ by extending the number system to include a new number, i (sometimes written as j). This has the property that $i^2 = -1$ and it follows the usual laws of algebra. i is called an **imaginary** number.

The square root of any negative number can be expressed in terms of i. For example, the solution of the equation $x^2 = -9$ is $x = \pm\sqrt{-9}$. This can be written as $\pm\sqrt{9} \times \sqrt{-1}$ which simplifies to $\pm 3i$.

| Example 2.1 | Use the quadratic formula to solve the quadratic equation $z^2 - 6z + 58 = 0$, simplifying your answer as far as possible. |

TECHNOLOGY

If your calculator has an equation solver, find out if it will give you the complex roots of this quadratic equation.

Solution

$z^2 - 6z + 58 = 0$ ← Using the quadratic formula with $a = 1$, $b = -6$ and $c = 58$.

$$z = \frac{6 \pm \sqrt{(-6)^2 - 4 \times 1 \times 58}}{2 \times 1}$$

$$= \frac{6 \pm \sqrt{-196}}{2}$$

$$= \frac{6 \pm 14i}{2}$$ ← $\sqrt{-196} = \sqrt{196} \times \sqrt{-1} = 14i$.

$$= 3 \pm 7i$$

3 is called the real part of the complex number $3 + 7i$ and is denoted $\operatorname{Re}(z)$.

You will have noticed that the roots $3 + 7i$ and $3 - 7i$ of the quadratic equation $z^2 - 6z + 58 = 0$ have both a **real part** and an **imaginary part**.

7 is called the imaginary part of the complex number and is denoted $\operatorname{Im}(z)$.

Notation

Any number z of the form $x + yi$, where x and y are real, is called a **complex number**.

The letter z is commonly used for complex numbers, and w is also used. In this chapter a complex number z is often denoted by $x + yi$, but other letters are sometimes used, such as $a + bi$.

x is called the real part of the complex number, denoted by $\operatorname{Re}(z)$ and y is called the imaginary part, denoted by $\operatorname{Im}(z)$.

Discussion points

→ What are the values of i^3, i^4, i^5, i^6 and i^7?

→ Explain how you could quickly work out the value of i^n for any positive integer value of n.

Working with complex numbers

The general methods for addition, subtraction and multiplication of complex numbers are straightforward.

Addition: add the real parts and add the imaginary parts.

For example, $(3 + 4i) + (2 - 8i) = (3 + 2) + (4 - 8)i$

$$= 5 - 4i$$

Subtraction: subtract the real parts and subtract the imaginary parts.

For example, $(6 - 9i) - (1 + 6i) = 5 - 15i$

Multiplication: multiply out the brackets in the usual way and simplify.

For example, $(7 + 2i)(3 - 4i) = 21 - 28i + 6i - 8i^2$

$$= 21 - 22i - 8(-1)$$

$$= 29 - 22i$$

> When simplifying it is important to remember that $i^2 = -1$.

Division of complex numbers follows later in this chapter.

Historical note

Gerolamo Cardano (1501–1576) was an Italian mathematician and physicist who was the first known writer to explore calculations involving the square roots of negative quantities, in his 1545 publication *Ars magna* ('The Great Art'). He wanted to calculate:

$$\left(5 + \sqrt{-15}\right)\left(5 - \sqrt{-15}\right)$$

Some years later, an Italian engineer named Rafael Bombelli introduced the words 'plus of minus' to indicate $\sqrt{-1}$ and 'minus of minus' to indicate $-\sqrt{-1}$.

However, the general mathematical community was slow to accept these 'fictional' numbers, with the French mathematician and philosopher René Descartes rather dismissively describing them as 'imaginary'. Similarly, Isaac Newton described the numbers as 'impossible' and the mystification of Gottfried Leibniz is evident in the quote at the beginning of the chapter! In the end it was Leonhard Euler who eventually began to use the symbol i, the first letter of 'imaginarius' (imaginary) instead of writing $\sqrt{-1}$.

Equality of complex numbers

Two complex numbers $z = x + yi$ and $w = u + vi$ are equal if both $x = u$ and $y = v$. If $x \neq u$ or $y \neq v$, or both, then z and w are not equal.

You may feel that this is obvious, but it is interesting to compare this situation with the equality of rational numbers.

> **Discussion points**
> → Are the rational numbers $\frac{x}{y}$ and $\frac{u}{v}$ equal if $x = u$ and $y = v$?
> → Is it possible for the rational numbers $\frac{x}{y}$ and $\frac{u}{v}$ to be equal if $x \neq u$ and $y \neq v$?

For two complex numbers to be equal the real parts must be equal and the imaginary parts must be equal. Using this result is described as **equating real** and **imaginary parts**, as shown in the following example.

Example 2.2

The complex numbers z_1 and z_2 are given by

$$z_1 = (3 - a) + (2b - 4)i$$

and

$$z_2 = (7b - 4) + (3a - 2)i.$$

(i) Given than z_1 and z_2 are equal, find the values of a and b.

(ii) Check your answer by substituting your values for a and b into the expressions above.

TECHNOLOGY

Some calculators will allow you to calculate with complex numbers. Find out whether your calculator has this facility.

Discussion point

→ What answer do you think Gerolamo Cardano might have obtained to the calculation $\left(5 + \sqrt{-15}\right)$ $\left(5 - \sqrt{-15}\right)$?

Solution

(i) $(3 - a) + (2b - 4)i = (7b - 4) + (3a - 2)i$

Equating real parts: $3 - a = 7b - 4$

> Equating real and imaginary parts leads to two equations.

Equating imaginary parts: $2b - 4 = 3a - 2$

$$\begin{cases} 7b + a = 7 \\ 2b - 3a = 2 \end{cases}$$

> Simplifying the equations.

Solving simultaneously gives $b = 1$ and $a = 0$.

(ii) Substituting $a = 0$ and $b = 1$ gives $z_1 = 3 - 2i$ and $z_2 = 3 - 2i$ so z_1 and z_2 are indeed equal.

Exercise 2.1

Do not use a calculator in this exercise

① Write down the values of

(i) i^9 (ii) i^{14} (iii) i^{31} (iv) i^{100}

② Find the following:

(i) $(6 + 4i) + (3 - 5i)$ (ii) $(-6 + 4i) + (-3 + 5i)$

(iii) $(6 + 4i) - (3 - 5i)$ (iv) $(-6 + 4i) - (-3 + 5i)$

③ Find the following:

(i) $3(6 + 4i) + 2(3 - 5i)$ (ii) $3i(6 + 4i) + 2i(3 - 5i)$

(iii) $(6 + 4i)^2$ (iv) $(6 + 4i)(3 - 5i)$

④ (i) Find the following:

(a) $(6 + 4i)(6 - 4i)$

(b) $(3 - 5i)(3 + 5i)$

(c) $(6 + 4i)(6 - 4i)(3 - 5i)(3 + 5i)$

(ii) What do you notice about the answers in part (i)?

⑤ Find the following:

(i) $(3 - 7i)(2 + 2i)(5 - i)$ (ii) $(3 - 7i)^3$

⑥ Solve each of the following equations.
In each case, check your solutions are correct by substituting the values back into the equation.

(i) $z^2 + 2z + 2 = 0$ (ii) $z^2 - 2z + 5 = 0$

(iii) $z^2 - 4z + 13 = 0$ (iv) $z^2 + 6z + 34 = 0$

(v) $4z^2 - 4z + 17 = 0$ (vi) $z^2 + 4z + 6 = 0$

⑦ Given that the complex numbers

$z_1 = a^2 + (3 + 2b)i$

$z_2 = (5a - 4) + b^2 i$

are equal, find the possible values of a and b.

Hence list the possible pairs of complex numbers z_1 and z_2.

⑧ A complex number $z = a + b$i, where a and b are real, is squared to give an answer of $-16 + 30$i. Find the possible values of a and b.

⑨ Find the square roots of the complex number $-40 + 42$i.

⑩ Figure 2.2 shows the graph of $y = x^2 - 4x + 3$.

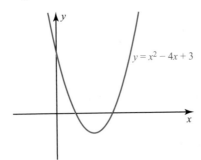

Figure 2.2

(i) Draw sketches of the curves $y = x^2 - 4x + 3$, $y = x^2 - 4x + 6$ and $y = x^2 - 4x + 8$ on the same axes.

(ii) Solve the equations

(a) $x^2 - 4x + 3 = 0$

(b) $x^2 - 4x + 6 = 0$

(c) $x^2 - 4x + 8 = 0$

(iii) Describe the relationship between the roots of the three equation and how they relate to the graphs you sketched in part (i).

⑪ Given that $z = 2 + 3$i is a root of the equation

$z^2 + (a - \text{i})z + 16 + b\text{i} = 0$

where a and b are real, find a and b.

Explain why you cannot assume that the other root is $z = 2 - 3$i.

Given that the second root has the form $5 + a\text{i}$, find the other root of the equation.

2 Division of complex numbers

Complex conjugates

You have seen that the roots of a quadratic equation with real coefficients are almost the same, but have the opposite sign (+ and −) between the real and imaginary terms. For example, the roots of $x^2 - 4x + 13 = 0$ are $x = 2 + 3$i and $x = 2 - 3$i. The pair of complex numbers $2 + 3$i and $2 - 3$i are called **conjugates**. Each is the conjugate of the other.

In general the complex number $x - y$i is called the **complex conjugate**, or just the conjugate, of $x + y$i. The conjugate of a complex number z is denoted by z^*.

Example 2.3

Given that $z = 3 + 5i$, find

(i) $z + z^*$ (ii) zz^*

Solution

(i) $z + z^* = (3 + 5i) + (3 - 5i)$
$$= 6$$

(ii) $zz^* = (3 + 5i)(3 - 5i)$
$$= 9 + 15i - 15i - 25i^2$$
$$= 9 + 25$$
$$= 34$$

ACTIVITY 2.3

Prove that $z + z^*$ and zz are both real for all complex numbers z.

You can see from the example above that $z + z^*$ and zz^* are both real. This is an example of an important general result: that the sum of two complex conjugates is real and that their product is also real.

Dividing complex numbers

You probably already know that you can write an expression like $\dfrac{2}{3 - \sqrt{2}}$ as a fraction with a rational denominator by multiplying the numerator and denominator by $3 + \sqrt{2}$.

$$\frac{2}{3 - \sqrt{2}} = \frac{2}{3 - \sqrt{2}} \times \frac{3 + \sqrt{2}}{3 + \sqrt{2}} = \frac{6 + 2\sqrt{2}}{9 - 2} = \frac{6 + 2\sqrt{2}}{7}$$

Because zz^* is always real, you can use a similar method to write an expression like $\dfrac{2}{3 - 5i}$ as a fraction with a real denominator, by multiplying the numerator and denominator by $3 + 5i$. ← | 3 + 5i is the complex conjugate of 3 − 5i.

This is the basis for dividing one complex number by another.

Example 2.4

Find the real and imaginary parts of $\dfrac{1}{5 + 2i}$.

Solution

Multiply the numerator and denominator by $5 - 2i$. ← | 5 − 2i is the conjugate of the denominator 5 + 2i.

$$\frac{1}{5 + 2i} = \frac{5 - 2i}{(5 + 2i)(5 - 2i)}$$
$$= \frac{5 - 2i}{25 + 4}$$
$$= \frac{5 - 2i}{29}$$

The real part is $\dfrac{5}{29}$ and the imaginary part is $-\dfrac{2}{29}$.

Example 2.5

Solve the equation $(2 + 3i)z = 9 - 4i$.

Discussion points

➜ What are the values of $\frac{1}{i}$, $\frac{1}{i^2}$, $\frac{1}{i^3}$ and $\frac{1}{i^4}$?

➜ Explain how you would work out the value of $\frac{1}{i^n}$ for any positive integer value n.

Solution

$(2 + 3i)z = 9 - 4i$

$\Rightarrow z = \dfrac{9 - 4i}{2 + 3i}$

$= \dfrac{(9 - 4i)(2 - 3i)}{(2 + 3i)(2 - 3i)}$

$= \dfrac{18 - 27i - 8i + 12i^2}{4 - 6i + 6i - 9i^2}$

$= \dfrac{6 - 35i}{13}$

$= \dfrac{6}{13} - \dfrac{35}{13}i$

> Multiply numerator and denominator by $2 - 3i$.

> Notice how the $-6i$ and $+6i$ terms will cancel to produce a real denominator.

Exercise 2.2

📈 TECHNOLOGY

If your calculator handles complex numbers, you can use it to check your answers.

① Express these complex numbers in the form $x + yi$.

(i) $\dfrac{3}{7 - i}$ (ii) $\dfrac{3}{7 + i}$ (iii) $\dfrac{3i}{7 - i}$ (iv) $\dfrac{3i}{7 + i}$

② Express these complex numbers in the form $x + yi$.

(i) $\dfrac{3 + 5i}{2 - 3i}$ (ii) $\dfrac{2 - 3i}{3 + 5i}$ (iii) $\dfrac{3 - 5i}{2 + 3i}$ (iv) $\dfrac{2 + 3i}{3 - 5i}$

③ Simplify the following, giving your answers in the form $x + yi$.

(i) $\dfrac{(12 - 5i)(2 + 2i)}{4 - 3i}$

(ii) $\dfrac{12 - 5i}{(4 - 3i)^2}$

④ $z = 3 - 6i$, $w = -2 + 9i$ and $q = 6 + 3i$.

Write down the values of the following:

(i) $z + z^*$ (ii) ww^* (iii) $q^* + q$

(iv) z^*z (v) $w + w^*$ (vi) qq^*

⑤ Given that $z = 2 + 3i$ and $w = 6 - 4i$, find the following:

(i) $\text{Re}(z)$ (ii) $\text{Im}(z)$ (iii) z^*

(iv) w^* (v) $z^* + w^*$ (vi) $z^* - w^*$

⑥ Given that $z = 2 + 3i$ and $w = 6 - 4i$, find the following:

(i) $\text{Im}(z + z^*)$ (ii) $\text{Re}(w - w^*)$ (iii) $zz^* - ww^*$

(iv) $(z^3)^*$ (v) $(z^*)^3$ (vi) $zw^* - z^*w$

⑦ Given that $z_1 = 2 - 5i$, $z_2 = 4 + 10i$ and $z_3 = 6 - 5i$, find the following in the form $a + bi$, where a and b are rational numbers.

(i) $\dfrac{z_1 z_2}{z_3}$ (ii) $\dfrac{(z_3)^2}{z_1}$ (iii) $\dfrac{z_1 + z_2 - z_3}{(z_3)^2}$

⑧ Solve these equations.

(i) $(1 + i)z = 3 + i$

(ii) $(2 - i)z + (2 - 6i) = 4 - 7i$

(iii) $(3 - 4i)(z - 1) = 10 - 5i$

(iv) $(3 + 5i)(z + 2 - 5i) = 6 + 3i$

⑨ Find the values of a and b such that $\dfrac{2 - 5i}{3 + 2i} = \dfrac{a + bi}{1 - i}$.

⑩ The complex number $w = a + bi$, where a and b are real, satisfies the equation $(5 - 2i)w = 67 + 37i$.

(i) Using the method of equating coefficients, find the values of a and b.

(ii) Using division of complex numbers, find the values of a and b.

⑪ (i) For $z = 5 - 8i$ find $\dfrac{1}{z} + \dfrac{1}{z^*}$ in its simplest form.

(ii) Write down the value of $\dfrac{1}{z} + \dfrac{1}{z^*}$ for $z = 5 + 8i$

⑫ For $z = x + yi$, find $\dfrac{1}{z} + \dfrac{1}{z^*}$ in terms of x and y.

⑬ Let $z_1 = x_1 + y_1 i$ and $z_2 = x_2 + y_2 i$.

Show that $(z_1 + z_2)^* = z_1^* + z_2^*$.

⑭ Find real numbers a and b such that $\dfrac{a}{3 + i} + \dfrac{b}{1 + 2i} = 1 - i$.

⑮ Find all the numbers z, real or complex, for which $z^2 = 2z^*$.

⑯ The complex numbers z and w satisfy the following simultaneous equations.

$z + wi = 13$

$3z - 4w = 2i$

Find z and w, giving your answers in the form $a + bi$.

3 Representing complex numbers geometrically

Discussion point

➜ Why is it not possible to show a complex number on a number line?

A complex number $x + yi$ can be represented by the point with Cartesian coordinates (x, y).

For example, in Figure 2.3,

$2 + 3i$ is represented by $(2, 3)$

$-5 - 4i$ is represented by $(-5, -4)$

$2i$ is represented by $(0, 2)$

7 is represented by $(7, 0)$.

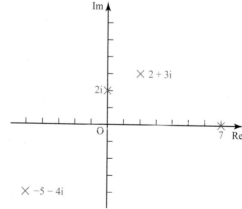

Figure 2.3

All real numbers are represented by points on the x-axis, which is therefore called the **real axis**. Purely imaginary numbers which have no real component (of the form $0 + y\mathrm{i}$) give points on the y-axis, which is called the **imaginary axis**.

These axes are labelled as Re and Im.

This geometrical illustration of complex numbers is called the **complex plane** or the **Argand diagram**. ◄———

> The Argand diagram is named after Jean-Robert Argand (1768–1822), a self-taught Swiss book-keeper who published an account of it in 1806.

ACTIVITY 2.4

(i) Copy Figure 2.3.

For each of the four given points z, mark also the point $-z$.
Describe the geometrical transformation which maps the point representing z to the point representing $-z$.

(ii) For each of the points z, mark the point z^*, the complex conjugate of z.
Describe the geometrical transformation which maps the point representing z to the point representing z^*.

Representing the sum and difference of complex numbers

In Figure 2.4 the complex number $z = x + y\mathrm{i}$ is shown as a vector on an Argand diagram.

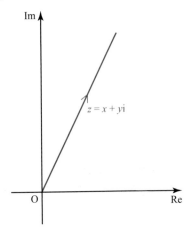

Figure 2.4

The use of vectors can be helpful in illustrating addition and subtraction of complex numbers on an Argand diagram. Figure 2.5 shows that the position vectors representing z_1 and z_2 form two sides of a parallelogram, the diagonal of which is the vector $z_1 + z_2$.

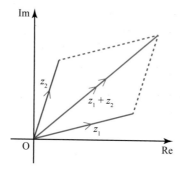

Figure 2.5

The addition can also be shown as a triangle of vectors, as in Figure 2.6.

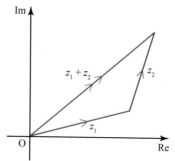

Figure 2.6

In Figure 2.7 you can see that $z_2 + w = z_1$ and so $w = z_1 - z_2$.

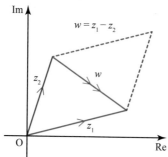

Figure 2.7

This shows that the complex number $z_1 - z_2$ is represented by the vector from the point representing z_2 to the point representing z_1, as shown in Figure 2.8.

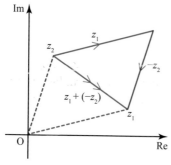

Figure 2.8

Notice the order of the points: the vector $z_1 - z_2$ starts at the point z_2 and goes to the point z_1.

Exercise 2.3

① Represent each of the following complex numbers on a single Argand diagram.

(i) $3 + 2i$ (ii) $4i$ (iii) $-5 + i$

(iv) -2 (v) $-6 - 5i$ (vi) $4 - 3i$

② Given that $z = 2 - 4i$, represent the following by points on a single Argand diagram.

(i) z (ii) $-z$ (iii) z^* (iv) $-z^*$

(v) iz (vi) $-iz$ (vii) iz^* (viii) $(iz)^*$

③ Given that $z = 10 + 5i$ and $w = 1 + 2i$, represent the following complex numbers on an Argand diagram.

(i) z (ii) w (iii) $z + w$

(iv) $z - w$ (v) $w - z$

④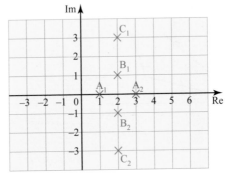

Figure 2.9

(i)   Find the quadratic equation which has roots A_1 and A_2.

(ii)   Find the quadratic equation which has roots B_1 and B_2.

(iii)   Find the quadratic equation which has roots C_1 and C_2.

(iv)   What do you notice about your answers to (i), (ii) and (iii)?

⑤ Give a geometrical proof that $(-z)^* = -(z^*)$.

⑥ Let $z = 1 + i$.

(i)   Find z^n for $n = -1, 0, 1, 2, 3, 4, 5$

(ii)   Plot each of the points z^n from part (i) on a single Argand diagram. Join each point to its predecessor and to the origin.

(iii)   Find the distance of each point from the origin.

(iv)   What do you notice?

⑦ Figure 2.10 shows the complex number $z = a + ib$. The distance of the point representing z from the origin is denoted by r.

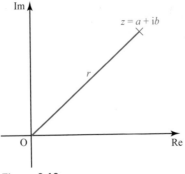

Figure 2.10

(i)   Find an expression for r, and hence prove that $r^2 = zz^*$.

A second complex number, w, is given by $w = c + d$i. The distance of the point representing w from the origin is denoted by s.

(ii)   Write down an expression for s.

(iii) Find zw, and prove that the distance of the point representing zw from the origin is given by rs.

LEARNING OUTCOMES

When you have completed this chapter you should be able to:

➤ understand how complex numbers extend the number system

➤ solve quadratic equations with complex roots

➤ know what is meant by the terms real part, imaginary part and complex conjugate

➤ add, subtract, multiply and divide complex numbers

➤ solve problems involving complex numbers by equating real and imaginary parts

➤ represent a complex number on an Argand diagram

➤ represent addition and subtraction of two complex numbers on an Argand diagram.

FUTURE USES

■ In Chapter 6 you will look at how complex numbers can be used to describe sets of points in the Argand diagram.

KEY POINTS

1 Complex numbers are of the form $z = x + y$i with $i^2 = -1$.
   x is called the real part, $\mathrm{Re}(z)$, and y is called the imaginary part, $\mathrm{Im}(z)$.

2 The conjugate of $z = x + y$i is $z^* = x - y$i .

3 To add or subtract complex numbers, add or subtract the real and imaginary parts separately.
   $$(x_1 + y_1 i) + (x_2 + y_2 i) = (x_1 + x_2) + (y_1 + y_2)i$$

4 To multiply complex numbers, expand the brackets then simplify using the fact that $i^2 = -1$

5 To divide complex numbers, write as a fraction, then multiply numerator and denominator by the conjugate of the denominator and simplify the answer.

6 Two complex numbers $z_1 = x_1 + y_1 i$ and $z_2 = x_2 + y_2 i$ are equal only if $x_1 = x_2$ and $y_1 = y_2$.

7 The complex number $z = x + y$i can be represented geometrically as the point (x, y).

   This is known as an Argand diagram. An Argand diagram is a way of representing complex numbers geometrically using Cartesian axes, the horizontal coordinate representing the real part of the number and the vertical coordinate the imaginary part.

3 Roots of polynomials

A **polynomial** is an expression like $4x^3 + x^2 - 4x - 1$. Its terms are all positive integer powers of a variable (in this case x) like x^2, or multiples of them like $4x^3$. There are no square roots, reciprocals, etc.

The **order** (or degree) of a polynomial is the highest power of the variable. So the order of $4x^3 + x^2 - 4x - 1$ is 3; this is why it is called a **cubic**.

You often need to solve polynomial equations, and it is usually helpful to think about the associated graph.

The following diagrams show the graphs of two cubic polynomial functions. The first example (in Figure 3.1) has three real roots (where the graph of the polynomial crosses the x-axis). The second example (in Figure 3.2) has only one real root. In this case there are also two **complex** roots.

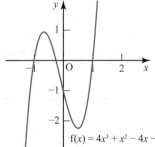

Figure 3.1 $f(x) = 4x^3 + x^2 - 4x - 1$

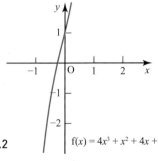

Figure 3.2 $f(x) = 4x^3 + x^2 + 4x + 1$

In general a polynomial equation of order n has n roots. However, some of these may be complex rather than real numbers and sometimes they coincide so that two distinct roots become one repeated root.

1 Polynomials

Discussion points

➜ How would you solve the polynomial equation $4x^3 + x^2 - 4x - 1 = 0$?

➜ What about $4x^3 + x^2 + 4x + 1 = 0$?

The following two statements are true for all polynomials:

- A polynomial equation of order n has at most n real roots.
- The graph of a polynomial function of order n has at most $n - 1$ turning points.

Here are some examples that illustrate these results.

Order 1 (a linear equation)

Example: $2x - 7 = 0$

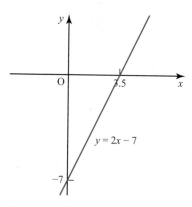

Figure 3.3 The graph is a straight line with no turning points. There is one real root at $x = 3.5$.

Order 3 (a cubic equation)

Example: $x^3 - 1 = 0$

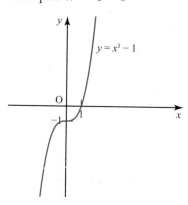

Figure 3.5 The two turning points of this curve coincide to give a point of inflection at $(0, -1)$. There is one real root at $x = 1$ and two complex roots at $x = \dfrac{-1 \pm \sqrt{3}i}{2}$.

The same patterns continue for higher order polynomials.

Order 2 (a quadratic equation)

Example: $x^2 - 4x + 4 = 0$

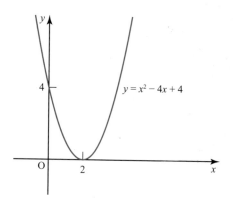

Figure 3.4 The curve has one turning point. There is one repeated root at $x = 2$.

Order 4 (a quartic equation)

Example: $x^4 - 3x^2 - 4 = 0$

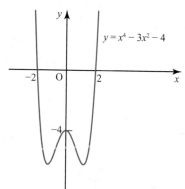

Figure 3.6 This curve has three turning points. There are two real roots at $x = -2$ and $x = 2$ and two complex roots at $x = \pm i$.

You will learn how to find the complex roots of polynomial equations later in this chapter.

The rest of this chapter explores some properties of polynomials, and ways to use these properties to avoid the difficulties of actually finding the roots of polynomials directly.

It is important that you recognise that the roots of polynomials may be complex. For this reason, in the work that follows, z is used as the variable (or unknown) instead of x to emphasise that the results apply regardless of whether the roots are complex or real.

Quadratic equations

ACTIVITY 3.1

TECHNOLOGY

You could use the equation solver on a calculator.

Solve each of the following quadratic equations (by factorising or otherwise). Also write down the *sum* and *product* of the two roots.

What do you notice?

Equation	Two roots	Sum of roots	Product of roots
(i) $z^2 - 3z + 2 = 0$			
(ii) $z^2 + z - 6 = 0$			
(iii) $z^2 - 6z + 8 = 0$			
(iv) $z^2 - 3z - 10 = 0$			
(v) $2z^2 - 3z + 1 = 0$			
(vi) $z^2 - 4z + 5 = 0$			

Discussion point

➙ What is the connection between the sums and products of the roots, and the coefficients in the original equation?

The roots of polynomial equations are usually denoted by Greek letters such as α and β. ⟵ α (alpha) and β (beta) are the first two letters of the Greek alphabet.

 Always be careful to distinguish between:
a – the coefficient of z^2 and
α – one of the roots of the quadratic.

If you know the roots are α and β, you can write the equation

$$az^2 + bz + c = 0$$

in factorised form as

$$a(z - \alpha)(z - \beta) = 0. \quad \longleftarrow \boxed{\text{Assuming } a \neq 0}$$

This gives the identity,

$$az^2 + bz + c \equiv a(z - \alpha)(z - \beta).$$

$$az^2 + bz + c \equiv a(z^2 - \alpha z - \beta z + \alpha\beta)$$

$$\equiv az^2 - a(\alpha + \beta)z + a\alpha\beta \quad \longleftarrow \boxed{\text{Multiplying out}}$$

Discussion point

→ What happens if you try to find the values of α and β by solving the equations $\alpha + \beta = -\dfrac{b}{a}$ and $\alpha\beta = \dfrac{c}{a}$ as a pair of simultaneous equations?

$$b = -a(\alpha + \beta) \Rightarrow \alpha + \beta = -\frac{b}{a}$$

Equating coefficients of z

$$c = a\alpha\beta \Rightarrow \alpha\beta = \frac{c}{a}$$

Equating constant terms

So the sum of the roots is

$$\alpha + \beta = -\frac{b}{a}$$

and the product of the roots is

$$\alpha\beta = \frac{c}{a}.$$

From these results you can obtain information about the roots without actually solving the equation.

ACTIVITY 3.2

The quadratic formula gives the roots of the quadratic equation $az^2 + bz + c = 0$ as

$$\alpha = \frac{-b + \sqrt{b^2 - 4ac}}{2a}, \qquad \beta = \frac{-b - \sqrt{b^2 - 4ac}}{2a}.$$

Use these expressions to prove that $\alpha + \beta = -\dfrac{b}{a}$ and $\alpha\beta = \dfrac{c}{a}$.

Example 3.1

Find a quadratic equation with roots 5 and -3.

Solution

The sum of the roots is $5 + (-3) = 2 \qquad \Rightarrow -\dfrac{b}{a} = 2$

The product of the roots is $5 \times (-3) = -15 \Rightarrow \dfrac{c}{a} = -15$

Taking a to be 1 gives ←

You could choose any value for a but choosing 1 in this case gives the simplest form of the equation.

$b = -2$ and $c = -15$

A quadratic equation with roots 5 and -3 is $z^2 - 2z - 15 = 0$.

Forming new equations

Using these properties of the roots sometimes allows you to form a new equation with roots that are related to the roots of the original equation. The next example illustrates this.

Example 3.2

The roots of the equation $2z^2 + 3z + 5 = 0$ are α and β.

(i) Find the values of $\alpha + \beta$ and $\alpha\beta$.

(ii) Find the quadratic equation with roots 2α and 2β.

Solution

(i) $\alpha + \beta = -\dfrac{3}{2}$ and

$\alpha\beta = \dfrac{5}{2}$

> These lines come from looking at the original quadratic, and quoting the facts $\alpha + \beta = -\dfrac{b}{a}$ and $\alpha\beta = \dfrac{c}{a}$.
> It might be confusing to introduce a, b and c here, since you need different values for them later in the question.

(ii) The sum of the new roots $= 2\alpha + 2\beta$

$= 2(\alpha + \beta)$

$= 2 \times -\dfrac{3}{2}$

$= -3$

The product of the new roots $= 2\alpha \times 2\beta$

$= 4\alpha\beta$

$= 4 \times \dfrac{5}{2}$

$= 10$

Let a, b and c be the coefficients in the new quadratic equation, then

$-\dfrac{b}{a} = -3$ and $\dfrac{c}{a} = 10$.

Taking $a = 1$ gives $b = 3$ and $c = 10$.

So a quadratic equation with the required roots is $z^2 + 3z + 10 = 0$.

Example 3.3

The roots of the equation $3z^2 - 4z - 1 = 0$ are α and β.
Find the quadratic equation with roots $\alpha + 1$ and $\beta + 1$.

Solution

$\alpha + \beta = \dfrac{4}{3}$ and

$\alpha\beta = -\dfrac{1}{3}$

The sum of the new roots $= \alpha + 1 + \beta + 1$

$= \alpha + \beta + 2$

$= \dfrac{4}{3} + 2$

$= \dfrac{10}{3}$

The product of the new roots $= (\alpha + 1)(\beta + 1)$
$$= \alpha\beta + (\alpha + \beta) + 1$$
$$= -\frac{1}{3} + \frac{4}{3} + 1$$
$$= 2$$

So $-\dfrac{b}{a} = \dfrac{10}{3}$ and $\dfrac{c}{a} = 2$.

Choose $a = 3$, then $b = -10$ and $c = 6$. ⟵

> Choosing $a = 1$ would give a value for b which is not an integer. It is easier here to use $a = 3$.

So a quadratic equation with the required roots is $3z^2 - 10z + 6 = 0$.

T

ACTIVITY 3.3

Solve the quadratic equations from the previous two examples (perhaps using the equation solver on your calculator, or a computer algebra system):

$2z^2 + 3z + 5 = 0$ $\qquad$ $z^2 + 3z + 10 = 0$

$3z^2 - 4z - 1 = 0$ $\qquad$ $3z^2 - 10z + 6 = 0$

Verify that the relationships between the roots are correct.

Exercise 3.1

① Write down the sum and product of the roots of each of these quadratic equations.

   (i) $\quad 2z^2 + 7z + 6 = 0$ $\qquad\qquad$ (ii) $\quad 5z^2 - z - 1 = 0$

   (iii) $\quad 7z^2 + 2 = 0$ $\qquad\qquad$ (iv) $\quad 5z^2 + 24z = 0$

   (v) $\quad z(z + 8) = 4 - 3z$ $\qquad\qquad$ (vi) $\quad 3z^2 + 8z - 6 = 0$

② Write down quadratic equations (in expanded form, with integer coefficients) with the following roots:

   (i) $\quad 7, 3$ $\qquad\qquad$ (ii) $\quad 4, -1$

   (iii) $\quad -5, -4.5$ $\qquad\qquad$ (iv) $\quad 5, 0$

   (v) $\quad 3$ (repeated) $\qquad\qquad$ (vi) $\quad 3 - 2i, 3 + 2i$

③ The roots of $2z^2 + 5z - 9 = 0$ are α and β.

Find quadratic equations with these roots.

   (i) $\quad 3\alpha$ and 3β $\qquad\qquad$ (ii) $\quad -\alpha$ and $-\beta$

   (iii) $\quad \alpha - 2$ and $\beta - 2$ $\qquad\qquad$ (iv) $\quad 1 - 2\alpha$ and $1 - 2\beta$

④ Using the fact that $\alpha + \beta = -\dfrac{b}{a}$, and $\alpha\beta = \dfrac{c}{a}$, what can you say about the roots, α and β, of $az^2 + bz + c = 0$ in the following cases:

   (i) $\quad a, b, c$ are all positive and $b^2 - 4ac > 0$

   (ii) $\quad b = 0$

   (iii) $\quad c = 0$

   (iv) $\quad a$ and c have opposite signs

⑤ One root of $az^2 + bz + c = 0$ is twice the other. Prove that $2b^2 = 9ac$.

⑥ The roots of $az^2 + bz + c = 0$ are, α and β. Find quadratic equations with the following roots:

> You may wish to introduce different letters (say p, q and r instead of a, b and c) for the coefficients of your target equation.

(i) $k\alpha$ and $k\beta$

(ii) $k + \alpha$ and $k + \beta$

⑦ (i) A quadratic equation with *real* coefficients $ax^2 + bx + c = 0$ has complex roots z_1 and z_2. Explain how the relationships between roots and coefficients show that z_1 and z_2 must be complex conjugates.

(ii) Find a quadratic equation with *complex* coefficients which has roots $2 + 3i$ and $3 - i$.

2 Cubic equations

There are corresponding properties for the roots of higher order polynomials.

To see how to generalise the properties you can begin with the cubics in a similar manner to the discussion of the quadratics. As before, it is conventional to use Greek letters to represent the three roots: α, β and γ (gamma, the third letter of the Greek alphabet).

You can write the general cubic as
$$az^3 + bz^2 + cz + d = 0$$
or in factorised form as
$$a(z - \alpha)(z - \beta)(z - \gamma) = 0.$$
This gives the identity
$$az^3 + bz^2 + cz + d \equiv a(z - \alpha)(z - \beta)(z - \gamma).$$

> Check this for yourself.

Multiplying out the right-hand side gives
$$az^3 + bz^2 + cz + d \equiv az^3 - a(\alpha + \beta + \gamma)z^2 + a(\alpha\beta + \beta\gamma + \gamma\alpha)z - a\alpha\beta\gamma.$$

Comparing coefficients of z^2:
$$b = -a(\alpha + \beta + \gamma) \Rightarrow \alpha + \beta + \gamma = -\frac{b}{a} \quad \text{Sum of the roots:} \quad \Sigma\alpha$$

Comparing coefficients of z:
$$c = a(\alpha\beta + \beta\gamma + \gamma\alpha) \Rightarrow \alpha\beta + \beta\gamma + \gamma\alpha = \frac{c}{a} \quad \text{Sum of products of pairs of roots:} \quad \Sigma\alpha\beta$$

Comparing constant terms:
$$d = -a\alpha\beta\gamma \Rightarrow \alpha\beta\gamma = -\frac{d}{a} \quad \text{Product of the three roots:} \quad \Sigma\alpha\beta\gamma$$

Note

Notation

It often becomes tedious writing out the sums of various combinations of roots, so shorthand notation is often used:

$\sum \alpha = \alpha + \beta + \gamma$ the sum of individual roots (however many there are)

$\sum \alpha\beta = \alpha\beta + \beta\gamma + \gamma\alpha$ the sum of the products of pairs of roots

$\sum \alpha\beta\gamma = \alpha\beta\gamma$ the sum of the products of triples of roots (in this case only one)

Provided you know the degree of the equation (e.g. cubic, quartic, etc,) it will be quite clear what this means. Functions like these are called symmetric functions of the roots, since exchanging any two of α, β, γ will not change the value of the function.

Using this notation you can shorten tediously long expressions. For example, for a cubic with roots α, β and γ,

$$\alpha^2\beta + \alpha\beta^2 + \beta^2\gamma + \beta\gamma^2 + \gamma^2\alpha + \gamma\alpha^2 = \sum \alpha^2\beta.$$

This becomes particularly useful when you deal with quartics in the next section.

Example 3.4

The roots of the equation $2z^3 - 9z^2 - 27z + 54 = 0$ form a geometric progression (i.e. they may be written as $\frac{a}{r}$, a, ar).

Solve the equation.

Solution

$\sum \alpha\beta\gamma = -\dfrac{d}{a}$ $\Rightarrow \dfrac{a}{r} \times a \times ar = -\dfrac{54}{2}$

                        $\Rightarrow a^3 = -27$

                        $\Rightarrow a = -3$

$\sum \alpha = -\dfrac{b}{a}$ $\Rightarrow \dfrac{a}{r} + a + ar = \dfrac{9}{2}$

                        $\Rightarrow -3\left(\dfrac{1}{r} + 1 + r\right) = \dfrac{9}{2}$

                        $\Rightarrow 2\left(\dfrac{1}{r} + 1 + r\right) = -3$

                        $\Rightarrow 2 + 2r + 2r^2 = -3r$

                        $\Rightarrow 2r^2 + 5r + 2 = 0$

                        $\Rightarrow (2r + 1)(r + 2) = 0$

                        $\Rightarrow r = -2$ or $r = -\dfrac{1}{2}$

Either value of r gives three roots: $\dfrac{3}{2}, -3, 6$.

Forming new equations: the substitution method

In the next example you are asked to form a new cubic equation with roots related to the roots of the original equation. Using the same approach as in the quadratic example is possible, but this gets increasingly complicated as the order of the equation increases. A substitution method is often a quicker alternative. The following example shows both methods for comparison.

Example 3.5	The roots of the cubic equation $2z^3 + 5z^2 - 3z - 2 = 0$ are α, β, γ.

Find the cubic equation with roots $2\alpha + 1, 2\beta + 1, 2\gamma + 1$.

Solution 1

$$\sum \alpha = \alpha + \beta + \gamma = -\frac{5}{2} \qquad \boxed{\sum \alpha = -\frac{b}{a}}$$

$$\sum \alpha\beta = \alpha\beta + \beta\gamma + \gamma\alpha = -\frac{3}{2} \qquad \sum \alpha\beta = \frac{c}{a}$$

$$\sum \alpha\beta\gamma = \alpha\beta\gamma = \frac{2}{2} = 1 \qquad \sum \alpha\beta\gamma = -\frac{d}{a}$$

For the new equation:

Sum of roots $= 2\alpha + 1 + 2\beta + 1 + 2\gamma + 1$

$$= 2(\alpha + \beta + \gamma) + 3$$

$$= -5 + 3 = -2$$

Product of the roots in pairs

$$= (2\alpha + 1)(2\beta + 1) + (2\beta + 1)(2\gamma + 1) + (2\gamma + 1)(2\alpha + 1)$$

$$= [4\alpha\beta + 2(\alpha + \beta) + 1] + [4\beta\gamma + 2(\beta + \gamma) + 1] + [4\gamma\alpha + 2(\gamma + \alpha) + 1]$$

$$= 4(\alpha\beta + \beta\gamma + \gamma\alpha) + 4(\alpha + \beta + \gamma) + 3$$

$$= 4 \times -\frac{3}{2} + 4 \times -\frac{5}{2} + 3$$

$$= -13$$

Product of roots $= (2\alpha + 1)(2\beta + 1))(2\gamma + 1)$

$$= 8\alpha\beta\gamma + 4(\alpha\beta + \beta\gamma + \gamma\alpha) + 2(\alpha + \beta + \gamma) + 1$$

$$= 8 \times 1 + 4 \times -\frac{3}{2} + 2 \times -\frac{5}{2} + 1$$

$$= -2$$

> Check this for yourself.

In the new equation, $-\frac{b}{a} = -2, \frac{c}{a} = -13, -\frac{d}{a} = -2$.

The new equation is $z^3 + 2z^2 - 13z + 2 = 0$.

> These are all integers, so choose $a = 1$ and this gives the simplest integer coefficients.

Solution 2 (substitution method)

> This is a transformation of z in the same way as the new roots are a transformation of the original z roots.

This method involves a new variable $w = 2z + 1$. You write z in terms of w, and substitute into the original equation:

$$z = \frac{w - 1}{2}$$

α, β, γ are the roots of $2z^3 + 5z^2 - 3z - 2 = 0$

$\Leftrightarrow$ $2\alpha + 1, 2\beta + 1, 2\gamma + 1$ are the roots of

$$2\left(\frac{w - 1}{2}\right)^3 + 5\left(\frac{w - 1}{2}\right)^2 - 3\left(\frac{w - 1}{2}\right) - 2 = 0$$

TECHNOLOGY

Use graphing software to draw the graphs of $y = 2x^3 + 5x^2 - 3x - 2$ and $y = x^3 + 2x^2 - 13x + 2$. How do these graphs relate to Example 3.5? What transformations map the first graph on to the second one?

$$\Leftrightarrow \frac{2}{8}(w-1)^3 + \frac{5}{4}(w-1)^2 - \frac{3}{2}(w-1) - 2 = 0$$

$$\Leftrightarrow (w-1)^3 + 5(w-1)^2 - 6(w-1) - 8 = 0$$

$$\Leftrightarrow w^3 - 3w^2 + 3w - 1 + 5w^2 - 10w + 5 - 6w + 6 - 8 = 0$$

$$\Leftrightarrow w^3 + 2w^2 - 13w + 2 = 0$$

The substitution method can sometimes be much more efficient, although you need to take care with the expansion of the cubic brackets.

Exercise 3.2

① The roots of the cubic equation $2z^3 + 3z^2 - z + 7 = 0$ are α, β, γ. Find the following:

  (i) $\sum \alpha$ (ii) $\sum \alpha\beta$ (iii) $\sum \alpha\beta\gamma$

② Find cubic equations (with integer coefficients) with the following roots:
  (i) $1, 2, 4$
  (ii) $2, -2, 3$
  (iii) $0, -2, -1.5$
  (iv) 2 (repeated), 2.5
  (v) $-2, -3, 5$
  (vi) $1, 2+i, 2-i$

③ The roots of each of these equations are in arithmetic progression (i.e. they may be written as $a - d$, a, $a + d$).
  Solve each equation.
  (i) $z^3 - 15z^2 + 66z - 80 = 0$
  (ii) $9z^3 - 18z^2 - 4z + 8 = 0$
  (iii) $z^3 - 6z^2 + 16 = 0$
  (iv) $54z^3 - 189z^2 + 207z - 70 = 0$

④ The roots of the equation $z^3 + z^2 + 2z - 3 = 0$ are α, β, γ.
  (i) The substitution $w = z + 3$ is made. Write z in terms of w.
  (ii) Substitute your answer to part (i) for z in the equation $z^3 + z^2 + 2z - 3 = 0$
  (iii) Give your answer to part (ii) as a cubic equation in w with integer coefficients.
  (iv) Write down the roots of your equation in part (iii), in terms of α, β and γ.

⑤ The roots of the equation $z^3 - 2z^2 + z - 3 = 0$ are α, β, γ. Use the substitution $w = 2z$ to find a cubic equation in w with roots $2\alpha, 2\beta, 2\gamma$.

⑥ The roots of the equation $2z^3 + 4z^2 - 3z + 1 = 0$ are α, β, γ.
  Find cubic equations with these roots:
  (i) $2 - \alpha, 2 - \beta, 2 - \gamma$
  (ii) $3\alpha - 2, 3\beta - 2, 3\gamma - 2$

⑦ The roots of the equation $2z^3 - 12z^2 + kz - 15 = 0$ are in arithmetic progression.
  Solve the equation and find k.

⑧ Solve $32z^3 - 14z + 3 = 0$ given that one root is twice another.

⑨ The equation $z^3 + pz^2 + 2pz + q = 0$ has roots $\alpha, 2\alpha, 4\alpha$.
  Find all possible values of p, q, α.

⑩ The roots of $z^3 + pz^2 + qz + r = 0$ are $\alpha, -\alpha, \beta$, and $r \neq 0$.
  Show that $r = pq$, and find all three roots in terms of p and q.

⑪ The cubic equation $8x^3 + px^2 + qx + r = 0$ has roots α and $\frac{1}{2a}$ and β.

(i) Express p, q and r in terms of α and β.

(ii) Show that $2r^2 - pr + 4q = 16$.

(iii) Given that $p = 6$ and $q = -23$, find the two possible values of r and, in each case, solve the equation $8x^3 + 6x^2 - 23x + r = 0$.

⑫ Show that one root of $az^3 + bz^2 + cz + d = 0$ is the reciprocal of another root if and only if $a^2 - d^2 = ac - bd$.

Verify that this condition is satisfied for the equation

$21z^3 - 16z^2 - 95z + 42 = 0$ and hence solve the equation.

⑬ Find a formula connecting a, b, c and d which is a necessary and sufficient condition for the roots of the equation $az^3 + bz^2 + cz + d = 0$ to be in geometric progression.

Show that this condition is satisfied for the equation

$8z^3 - 52z^2 + 78z - 27 = 0$ and hence solve the equation.

3 Quartic equations

Quartic equations have four roots, denoted by the first four Greek letters: α, β, γ and δ (delta).

> ### Discussion point
>
> → By looking back at the two formulae for quadratics and the three formulae for cubics, predict the *four* formulae that relate the roots α, β, γ and δ to the coefficients a, b, c and d of the quartic equation $ax^4 + bx^3 + cx^2 + dx + e = 0$.
>
> You may wish to check/derive these results yourself before looking at the derivation on the next page.

 ### Historical note

The formulae used to relate the coefficients of polynomials with sums and products of their roots are called Vieta's Formulae after François Viète (a Frenchman who commonly used a Latin version of his name: Franciscus Vieta). He was a lawyer by trade but made important progress (while doing mathematics in his spare time) on algebraic notation and helped pave the way for the more logical system of notation you use today.

Derivation of formulae

As before, the quartic equation

$$az^4 + bz^3 + cz^2 + dz + e = 0$$

can be written is factorised form as

$$a(z - \alpha)(z - \beta)(z - \gamma)(z - \delta) = 0.$$

This gives the identity

$$az^4 + bz^3 + cz^2 + dz + e \equiv a(z - \alpha)(z - \beta)(z - \gamma)(z - \delta).$$

Multiplying out the right-hand side gives

$$az^4 + bz^3 + cz^2 + dz + e \equiv az^4 - a(\alpha + \beta + \gamma + \delta)z^3$$
$$+ a(\alpha\beta + \alpha\gamma + \alpha\delta + \beta\gamma + \beta\delta + \gamma\delta)z^2 - a(\alpha\beta\gamma + \beta\gamma\delta$$
$$+ \gamma\delta\alpha + \delta\alpha\beta)z + a\alpha\beta\gamma\delta.$$

Check this for yourself.

Equating coefficients shows that

$$\sum\alpha = \alpha + \beta + \gamma + \delta = -\frac{b}{a}$$

The sum of the individual roots.

$$\sum\alpha\beta = \alpha\beta + \alpha\gamma + \alpha\delta + \beta\gamma + \beta\delta + \gamma\delta = \frac{c}{a}$$

The sum of the products of roots in pairs.

$$\sum\alpha\beta\gamma = \alpha\beta\gamma + \beta\gamma\delta + \gamma\delta\alpha + \delta\alpha\beta = -\frac{d}{a}$$

The sum of the products of roots in threes.

$$\alpha\beta\gamma\delta = \frac{e}{a}$$

The product of the roots.

Example 3.6

The roots of the quartic equation $4z^4 + pz^3 + qz^2 - z + 3 = 0$ are $\alpha, -\alpha, \alpha + \lambda, \alpha - \lambda$ where α and λ are real numbers.

(i)   Express p and q in terms of α and λ.

(ii)   Show that $\alpha = -\frac{1}{2}$, and find the values of p and q.

(iii)   Give the roots of the quartic equation.

Solution

(i) $\sum\alpha = \alpha - \alpha + \alpha + \lambda + \alpha - \lambda = -\frac{p}{4}$

$$\Rightarrow 2\alpha = -\frac{p}{4}$$

$$\Rightarrow p = -8\alpha$$

Use the sum of the individual roots to find an expression for p.

$$\sum\alpha\beta = -\alpha^2 + \alpha(\alpha + \lambda) + \alpha(\alpha - \lambda) - \alpha(\alpha + \lambda) - \alpha(\alpha - \lambda)$$
$$+ (\alpha + \lambda)(\alpha - \lambda) = \frac{q}{4}$$

$$\Rightarrow -\lambda^2 = \frac{q}{4}$$

$$\Rightarrow q = -4\lambda^2$$

Use the sum of the product of the roots in pairs to find an expression for q.

(ii)
$$\sum\alpha\beta\gamma = -\alpha^2(\alpha + \lambda) - \alpha(\alpha + \lambda)(\alpha - \lambda) + \alpha(\alpha + \lambda)(\alpha - \lambda) - \alpha^2(\alpha - \lambda) = \frac{1}{4}$$

$$\Rightarrow -2\alpha^3 = \frac{1}{4}$$

$$\Rightarrow \alpha = -\frac{1}{2}$$

$$p = -8\alpha = -8 \times -\frac{1}{2} = 4$$

Use the sum of the product of the roots in threes to find α (λ cancels out) and hence find p, using your answer to part (i).

$$\alpha\beta\gamma\delta = -\alpha^2(\alpha + \lambda)(\alpha - \lambda) = \frac{3}{4}$$

> Use the sum of the product of the roots and the value for α to find λ, and hence find q, using your answer to part (i).

$$\Rightarrow -\alpha^2(\alpha^2 - \lambda^2) = \frac{3}{4}$$

$$\Rightarrow -\frac{1}{4}\left(\frac{1}{4} - \lambda^2\right) = \frac{3}{4}$$

$$\Rightarrow \frac{1}{4} - \lambda^2 = -3$$

$$\Rightarrow \lambda^2 = \frac{13}{4}$$

> Substitute the values for α and λ to give the roots.

$$q = -4\lambda^2 = -4 \times \frac{13}{4} = -13$$

(iii) The roots of the equation are $\frac{1}{2}, -\frac{1}{2}, -\frac{1}{2} + \frac{1}{2}\sqrt{13}, -\frac{1}{2} - \frac{1}{2}\sqrt{13}$.

Exercise 3.3

① The roots of $2z^4 + 3z^3 + 6z^2 - 5z + 4 = 0$ are α, β, γ and δ.

Write down the values of the following:

(i) $\sum \alpha$　　(ii) $\sum \alpha\beta$　　(iii) $\sum \alpha\beta\gamma$　　(iv) $\sum \alpha\beta\gamma\delta$

② Find quartic equations (with integer coefficients) with the roots.

(i) $1, -1, 2, 4$

(ii) $0, 1.5, -2.5, -4$

(iii) 1.5 (repeated), -3 (repeated)

(iv) $1, -3, 1 + i, 1 - i$.

③ The roots of the quartic equation $2z^4 + 4z^3 - 3z^2 - z + 6 = 0$ are α, β, γ and δ.

Find quartic equations with these roots:

(i) $2\alpha, 2\beta, 2\gamma, 2\delta$

(ii) $\alpha - 1, \beta - 1, \gamma - 1, \delta - 1$.

④ The roots of the quartic equation $x^4 + 4x^3 - 8x + 4 = 0$ are α, β, γ and δ.

(i) By making a suitable substitution, find a quartic equation with roots $\alpha + 1, \beta + 1, \gamma + 1$ and $\delta + 1$.

(ii) Solve the equation found in part (i), and hence find the values of α, β, γ and δ.

⑤ The quartic equation $x^4 + px^3 - 12x + q = 0$, where p and q are real, has roots $\alpha, 3\alpha, \beta, -\beta$.

(i) By considering the coefficients of x^2 and x, find α and β, where $\beta > 0$.

(ii) Show that $p = 4$ and find the value of q.

(iii) By making the substitution $y = x - k$, for a suitable value of k, find a **cubic** equation in y, with integer coefficients, which has roots

$-2\alpha, \beta - 3\alpha, -\beta - 3\alpha$.

⑥ (i) Make conjectures about the five properties of the roots $\alpha, \beta, \gamma, \delta$ and ε (epsilon) of the general quintic $ax^5 + bx^4 + cx^3 + dx^2 + ex + f = 0$.

(ii) Prove your conjectures.

4 Solving polynomial equations with complex roots

Prior Knowledge

You need to know how to use the factor theorem to solve polynomial equations (covered in AS Mathematics Chapter 7).

When solving polynomial equations with real coefficients, it is important to remember that any complex roots occur in conjugate pairs.

When there is a possibility of complex roots, it is common to express the polynomial in terms of z.

Example 3.7

The equation $z^3 + 7z^2 + 17z + 15 = 0$ has one integer root.

(i)   Factorise $f(z) = z^3 + 7z^2 + 17z + 15$.

(ii)   Solve $z^3 + 7z^2 + 17z + 15 = 0$.

(iii)   Sketch the graph of $y = x^3 + 7x^2 + 17x + 15$.

Solution

(i) $f(1) = 1^3 + 7 \times 1^2 + 17 \times 1 + 15 = 40$

> If there is an integer root, it must be a factor of 15. So try $z = \pm 1, \pm 3$, etc.

$f(-1) = (-1)^3 + 7 \times (-1)^2 + 17 \times (-1) + 15 = 4$

$f(3) = 3^3 + 7 \times 3^2 + 17 \times 3 + 15 = 156$

$f(-3) = (-3)^3 + 7 \times (-3)^2 + 17 \times (-3) + 15 = 0$

> $f(-3) = 0$ so using the factor theorem, $(z + 3)$ is a factor.

So one root is $z = -3$, and $(z + 3)$ is a factor of $f(z)$.

Using algebraic division or by inspection, $f(z)$ can be written in the form:

$f(z) = (z + 3)(z^2 + 4z + 5)$

> Check this for yourself

Now solve the quadratic equation $z^2 + 4z + 5 = 0$:

> using the quadratic formula.

$$z = \frac{-4 \pm \sqrt{4^2 - (4 \times 1 \times 5)}}{2} = \frac{-4 \pm \sqrt{-4}}{2} = \frac{-4 \pm 2i}{2} = -2 \pm i$$

So, fully factorised $f(z) = (z + 3)(z - (-2 + i))(z - (-2 - i))$

(ii) The roots are $z = -3$, $z = -2 \pm i$

> Note that there is one conjugate pair of complex roots and one real root.

(iii) Figure 3.7 shows the graph of the curve $y = f(x)$.

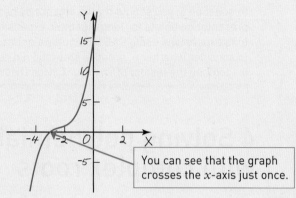

> You can see that the graph crosses the x-axis just once.

Figure 3.7

Sometimes you can use the relationships between roots and coefficients of polynomial equations to help you to find the roots. In Example 3.8, two solution methods are shown.

Example 3.8

Given that $z = 1 + 2i$ is a root of $4z^3 - 11z^2 + 26z - 15 = 0$, find the other roots.

Solution 1

As complex roots occur in conjugate pairs, the conjugate $z = 1 - 2i$ is also a root.

The next step is to find a quadratic equation $az^2 + bz + c = 0$ with roots $1 + 2i$ and $1 - 2i$.

$$-\frac{b}{a} = (1 + 2i) + (1 - 2i) = 2$$

$$\frac{c}{a} = (1 + 2i)(1 - 2i) = 1 + 4 = 5$$

Taking $a = 1$ gives $b = -2$ and $c = 5$
So the quadratic equation is $z^2 - 2z + 5 = 0$
$$4z^3 - 11z^2 + 26z - 15 = (z^2 - 2z + 5)(4z - 3)$$
The other roots are $z = 1 - 2i$ and $z = \frac{3}{4}$.

Solution 2

The sum of the three roots is $\frac{11}{4}$ ← $\boxed{\alpha + \beta + \gamma = -\frac{b}{a}}$

$$1 + 2i + 1 - 2i + \gamma = \frac{11}{4}$$

$$\gamma = \frac{11}{4} - 2 = \frac{3}{4}$$

The other roots are $z = 1 - 2i$ and $z = \frac{3}{4}$.

Notice that Solution 2 is more efficient than Solution 1 in this case. You should look out for situations like this where using the relationships between roots and coefficients can be helpful.

Example 3.9

(i)   Solve $z^4 - 3z^2 - 4 = 0$.

(ii)   Sketch the curve $y = x^4 - 3x^2 - 4$.

(iii)   Show the roots of $z^4 - 3z^2 - 4 = 0$ on an Argand diagram.

Solution

(i) $z^4 - 3z^2 - 4 = 0$
$(z^2 - 4)(z^2 + 1) = 0$ ◄──── $z^4 - 3z^2 - 4$ is a quadratic in z^2 and can be factorised.
$(z - 2)(z + 2)(z + i)(z - i) = 0$

The solution is $z = 2, -2, i, -i$.

(ii)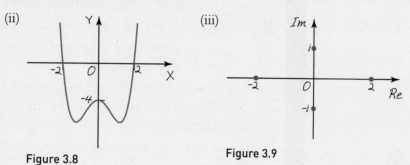

Figure 3.8

(iii) Figure 3.9

Exercise 3.4

① $4 - 5i$ is one root of a quadratic equation with real coefficients.
Write down the second root of the equation and hence find the equation.

② Verify that $2 + i$ is a root of $z^3 - z^2 - 7z + 15 = 0$, and find the other roots.

③ One root of $z^3 - 15z^2 + 76z - 140 = 0$ is an integer.
Solve the equation and show all three roots on an Argand diagram.

④ The equation $z^3 - 2z^2 - 6z + 27 = 0$ has a real integer root in the range $-6 \le z \le 0$.

(i)   Find the real root of the equation.

(ii)   Hence solve the equation and find the exact value of all three roots.

⑤ Given that 4 is a root of the equation
$z^3 - z^2 - 3z - k = 0$, find the value of k
and hence find the exact value of the other
two roots of the equation.

⑥ Given that $1 - i$ is a root of
$z^3 + pz^2 + qz + 12 = 0$, find the real
numbers p and q, and state the other roots.

⑦ The three roots of a cubic equation are shown
on the Argand diagram in Figure 3.10.

Find the equation in polynomial form.

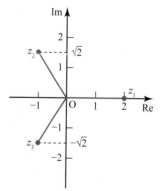

Figure 3.10

⑧ One root of $z^4 - 10z^3 + 42z^2 - 82z + 65 = 0$ is $3 + 2i$.

Solve the equation and show the four roots on an Argand diagram.

⑨ You are given the complex number $w = 1 - i$.

(i) Express w^2, w^3 and w^4 in the form $a + bi$.

(ii) Given that $w^4 + 3w^3 + pw^2 + qw + 8 = 0$, where p and q are real numbers, find the values of p and q.

(iii) Hence find all four roots of the equation $z^4 + 3z^3 + pz^2 + qz + 8 = 0$, where p and q are the real numbers found in part (ii).

⑩ (i) Solve the equation $z^4 - 81 = 0$

(ii) Hence show the four fourth roots of 81 on an Argand diagram.

⑪ (i) Given that $\alpha = -1 + 2i$, express α^2 and α^3 in the form $a + bi$.

Hence show that α is a root of the cubic equation
$z^3 + 7z^2 + 15z + 25 = 0$

(ii) Find the other two roots of this cubic equation.

(iii) Illustrate the three roots of the cubic equation on an Argand diagram.

⑫ For each of these statements about polynomial equations with real coefficients, say whether the statement is TRUE or FALSE, and give an explanation.

(A) A cubic equation can have three complex roots.

(B) Some equations of order 6 have no real roots.

(C) A cubic equation can have a single root repeated three times.

(D) A quartic equation can have a repeated complex root.

⑬ Given that $z = -2 + i$ is a root of the equation
$z^4 + az^3 + bz^2 + 10z + 25 = 0$, find the values of a and b, and solve the equation.

⑭ The equation $z^4 - 8z^3 + 20z^2 - 72z + 99 = 0$ has a purely imaginary root.

Solve the equation.

⑮ In this question, α is the complex number $-1 + 3i$.

(i) Find α^2 and α^3.

It is given that λ and μ are real numbers such that

$\lambda\alpha^3 + 8\alpha^2 + 34\alpha + \mu = 0$

(ii) Show that $\lambda = 3$, and find the value of μ.

(iii) Solve the equation $\lambda z^3 + 8z^2 + 34z + \mu = 0$, where λ and μ are as in part (ii).

(iv) Illustrate the three roots on an Argand diagram.

⑯ Three of the roots of the quintic equation $az^5 + bz^4 + cz^3 + dz^2 + ez + f = 0$ are $3, -4i$ and $3 - i$.

Find the values of the coefficients of the equation.

LEARNING OUTCOMES

When you have completed this chapter you should be able to:

➤ know the relationships between the roots and coefficients of quadratic, cubic and quartic equations

➤ form new equations whose roots are related to the roots of a given equation by a linear transformation

➤ understand that complex roots of polynomial equations with real coefficients occur in conjugate pairs

➤ solve cubic and quartic equations with complex roots.

KEY POINTS

1 If α and β are the roots of the quadratic equation $az^2 + bz + c = 0$, then
$$\alpha + \beta = -\frac{b}{a} \text{ and } \alpha\beta = \frac{c}{a}.$$

2 If α, β and γ are the roots of the cubic equation $az^3 + bz^2 + cz + d = 0$, then
$$\Sigma\alpha = \alpha + \beta + \gamma = -\frac{b}{a},$$
$$\Sigma\alpha\beta = \alpha\beta + \beta\gamma + \gamma\alpha = \frac{c}{a} \text{ and,}$$
$$\alpha\beta\gamma = -\frac{d}{a}.$$

3 If α, β, γ and δ are the roots of the quartic equation
$az^4 + bz^3 + cz^2 + dz + e = 0$, then
$$\Sigma\alpha = \alpha + \beta + \gamma + \delta = -\frac{b}{a},$$
$$\Sigma\alpha\beta = \alpha\beta + \alpha\gamma + \alpha\delta + \beta\gamma + \beta\delta + \gamma\delta = \frac{c}{a},$$
$$\Sigma\alpha\beta\gamma = \alpha\beta\gamma + \beta\gamma\delta + \gamma\delta\alpha + \delta\alpha\beta = -\frac{d}{a} \text{ and}$$
$$\alpha\beta\gamma\delta = \frac{e}{a}.$$

4 All of these formulae may be summarised using the shorthand sigma notation for elementary symmetric functions as follows:
$$\Sigma\alpha = -\frac{b}{a}$$
$$\Sigma\alpha\beta = \frac{c}{a}$$
$$\Sigma\alpha\beta\gamma = -\frac{d}{a}$$
$$\Sigma\alpha\beta\gamma\delta = \frac{e}{a}$$

(using the convention that polynomials of degree n are labelled
$az^n + bz^{n-1} + \ldots = 0$ and have roots α, β, γ, δ, ...)

5 A polynomial equation of degree n has n roots, taking into account complex roots and repeated roots. In the case of polynomial equations with real coefficients, complex roots always occur in conjugate pairs.

Figure 4.1

Discussion point

➜ How would you describe the sequence of pictures of the moon shown in Figure 4.1?

Discussion point

➜ How would you describe this sequence?

1 Sequences and series

A **sequence** is an ordered set of objects with an underlying rule.

For example:

$$2, 5, 8, 11, 14$$

A **series** is the sum of the terms of a numerical sequence:

$$2 + 5 + 8 + 11 + 14$$

Notation

There are a number of different notations which are commonly used in writing down sequences and series:

- The terms of a sequence are often written as $a_1, a_2, a_3, \ldots$ or $u_1, u_2, u_3, \ldots$
- The general term of a sequence may be written as a_r or u_r. (Sometimes the letters k or i are used instead of r.)
- The last term is usually written as a_n or u_n.
- The sum S_n of the first n terms of a sequence can be written using the symbol Σ (the Greek capital S, sigma).

$$S_n = a_1 + a_2 + a_3 + \ldots + a_n = \sum_{r=1}^{n} a_r$$

The numbers above and below the Σ are the limits of the sum. They show that the sum includes all the a_r from a_1 to a_n. The limits may be omitted if they are obvious, so that you would just write $\sum a_r$ or you might write $\sum_r a_r$ (meaning the sum of a_r for all values of r).

When discussing sequences you may find the following vocabulary helpful:

- In an **increasing sequence**, each term is greater than the previous term.
- In a **decreasing sequence**, each term is smaller than the previous term.
- In an **oscillating sequence**, the terms lie above and below a middle number.
- The terms of a **convergent sequence** get closer and closer to a limiting value.

Defining sequences

One way to define a sequence is by thinking about the relationship between one term and the next.

The sequence $2, 5, 8, 11, 14, \ldots$ can be written as

$$u_1 = 2 \quad \longleftarrow \boxed{\text{You need to say where the sequence starts.}}$$

$$u_{r+1} = u_r + 3 \quad \longleftarrow \boxed{\text{You find each term by adding 3 to the previous term.}}$$

This is called an **inductive** definition or **term-to-term** definition.

An alternative way to define a sequence is to describe the relationship between the term and its position.

In this case,

$$u_r = 3r - 1.$$

You can see that, for example, substituting $r = 2$ into this definition gives $u_2 = (3 \times 2) - 1 = 5$, which is the second term of the sequence.

This is called a **deductive** definition or **position-to-term** definition.

The series of positive integers

One of the simplest of all sequences is the sequence of the integers:

$$1, 2, 3, 4, 5, 6, \ldots$$

As simple as it is, it may not be immediately obvious how to calculate the sum of the first few integers, for example the sum of the first 100 integers.

$$\sum_{r=1}^{100} r = 1 + 2 + \ldots + 100$$

One way of reaching a total is illustrated below.

$$S_{100} = 1 + 2 + 3 + \ldots + 98 + 99 + 100 \quad \longleftarrow \quad \boxed{\text{Call the sum } S_{100}}$$

Rewrite S_{100} in reverse:

$$S_{100} = 100 + 99 + 98 + \ldots + 3 + 2 + 1$$

Adding these two lines together, by matching up each term with the one below it, produces pairings of 101 each time, while giving you $2S_{100}$ on the left-hand side.

$$
\begin{aligned}
S_{100} &= 1 + 2 + 3 + \ldots + 98 + 99 + 100 \\
S_{100} &= 100 + 99 + 98 + \ldots + 3 + 2 + 1 \\
\hline
2S_{100} &= 101 + 101 + 101 + \ldots + 101 + 101 + 101
\end{aligned}
$$

There are 100 terms on the right-hand side (since you were originally adding 100 terms together), so simplify the right-hand side:

$$2S_{100} = 100 \times 101$$

and solve for S_{100}:

$$2S_{100} = 10\,100$$

$$S_{100} = 5050$$

The sum of the first 100 integers is 5050.

You can use this method to find a general result for the sum of the first n integers (call this S_n).

$$S_n = 1 + 2 + 3 + \ldots + (n - 2) + (n - 1) + n$$

$$S_n = n + (n - 1) + (n - 2) + \ldots + 3 + 2 + 1$$

$$2S_n = (n + 1) + (n + 1) + (n + 1) + \ldots + (n + 1) + (n + 1) + (n + 1)$$

$$2S_n = n(n + 1)$$

$$S_n = \tfrac{1}{2}n(n + 1)$$

 **TECHNOLOGY**

You could use a spreadsheet to verify this result for different values of n.

This result is an important one and you will often need to use it.

Note

A common confusion occurs with the sigma notation when there is no r term present.

For example,

$$\sum_{r=1}^{5} 3 \longleftarrow \boxed{\text{This means 'The sum of 3, with } r \text{ changing from 1 to 5'.}}$$

means

$$3 + 3 + 3 + 3 + 3 = 15$$

since there are five terms in the sum (it's just that there is no r term to change anything each time).

In general:

$$\sum_{r=1}^{n} 1 = 1 + 1 + \ldots + 1 + 1$$

with n repetitions of the number 1.

So,

$$\sum_{r=1}^{n} 1 = n$$

This apparently obvious result is important and you will often need to use it.

You can use the results $\displaystyle\sum_{r=1}^{n} r = \frac{1}{2}n(n+1)$ and $\displaystyle\sum_{r=1}^{n} 1 = n$ to find the sum of other series.

Example 4.1

For the series $2 + 5 + 8 + \ldots + 500$:

(i)   Find a formula for the rth term, u_r.

(ii) How many terms are in this series?

(iii) Find the sum of the series using the reverse/add method.

(iv) Express the sum using sigma notation, and use this to confirm your answer to part (iii).

Solution

(i)   The terms increase by 3 each time and start at 2. So $u_r = 3r - 1$.

(ii) Let the number of terms be n. The last term (the nth term) is 500.

$$u_n = 3n - 1$$
$$3n - 1 = 500$$
$$3n = 501$$
$$n = 167$$

There are 167 terms in this series.

(iii) $S = 2 + 5 + \ldots + 497 + 500$

$S = 500 + 497 + \ldots + 5 + 2$

$2S = 167 \times 502$

$S = 41\,917$

(iv) $\quad S = \displaystyle\sum_{r=1}^{167}(3r - 1)$

$$S = \sum_{r=1}^{167}3r - \sum_{r=1}^{167}1$$

$$S = 3\sum_{r=1}^{167}r - \sum_{r=1}^{167}1$$

$$S = 3 \times \frac{1}{2} \times 167 \times 168 - 167$$

$$S = 41917$$

Using the results $\displaystyle\sum_{r=1}^{n}r = \frac{1}{2}n(n + 1)$

and $\displaystyle\sum_{r=1}^{n}1 = n$

Example 4.2

Calculate the sum of the integers from 100 to 200 inclusive.

Solution

$$\sum_{r=100}^{200}r = \sum_{1}^{200}r - \sum_{1}^{99}r$$

Start with all the integers from 1 to 200, and subtract the integers from 1 to 99, leaving those from 100 to 200.

$$= \frac{1}{2} \times 200 \times 201 - \frac{1}{2} \times 99 \times 100$$

$$= 20\,100 - 4950$$

$$= 15\,150$$

Exercise 4.1

① For each of the following definitions, write down the first five terms of the sequence and describe the sequence.

(i) $u_r = 5r + 1$

(ii) $v_r = 3 - 6r$

(iii) $p_r = 2^{r+2}$

(iv) $q_r = 10 + 2 \times (-1)^r$

(v) $a_{r+1} = 2a_r + 1, \quad a_1 = 2$

(vi) $u_r = \dfrac{5}{r}$

② For the sequence 1, 5, 9, 13, 17, …

(i)   write down the next four terms of the sequence

(ii)   write down an inductive rule for the sequence, in the form
$u_1 = \cdots, u_{r+1} = \cdots$

(iii) write down a deductive rule for the general term of the sequence, in the form $u_r = \cdots$

③ For each of the following sequences:

(a)   write down the next four terms of the sequence

(b)   write down an inductive rule for the sequence

(c)   write down a deductive rule for the general term of the sequence

(d)   find the 20th term of the sequence.

(i) $10, 8, 6, 4, 2, \ldots$

(ii) $1, 2, 4, 8, 16, \ldots$

(iii) $50, 250, 1250, 6250, \ldots$

④ Find the sum of the series $\displaystyle\sum_{1}^{5} u_r$ for each of the following:

(i) $u_r = 2 + r$

(ii) $u_r = 3 - 11r$

(iii) $u_r = 3^r$

(iv) $u_r = 7.5 \times (-1)^r$

⑤ For $S = 50 + 44 + 38 + 32 + \ldots + 14$

(i) Express S in the form $\displaystyle\sum_{r=1}^{n} u_r$

where n is an integer, and u_r is an algebraic expression for the rth term of the series.

(ii) Hence, or otherwise, calculate the value of S.

⑥ Given $u_r = 6r + 2$, calculate $\displaystyle\sum_{r=11}^{30} u_r$.

⑦ The general term of a sequence is given by $u_r = (-1)^r \times 5$.

(i) Write down the first six terms of the sequence and describe it.

(ii) Find the sum of the series $\displaystyle\sum_{r=1}^{n} u_r$:

(a) when n is even

(b) when n is odd.

(iii) Find an algebraic expression for the sum to n terms, whatever the value of n.

⑧ A sequence is given by

$b_{r+2} = b_r + 2, b_1 = 0, b_2 = 100$

(i) Write down the first six terms of the sequence and describe it.

(ii) Find the smallest odd value of r for which $b_r \geq 200$.

(iii) Find the largest even value of r for which $b_r \leq 200$.

⑨ A sawmill receives an order requesting many logs of various specific lengths, that must come from the same particular tree. The log lengths must start at 5 cm long and increase by 2 cm each time, up to a length of 53 cm.

The saw blade destroys 1 cm (in length) of wood (turning it to sawdust) at every cut. What is the minimum height of tree required to fulfil this order?

⑩ Find the sum of the integers from n to its square (inclusive). Express your answer in a fully factorised form.

⑪ Write down the first ten terms of the following sequence:

$$c_{r+1} = \begin{cases} 3c_r + 1 & \text{if } c_r \text{ is odd} \\ \dfrac{c_r}{2} & \text{if } c_r \text{ is even} \end{cases} \qquad c_1 = 10$$

You can find out more about this sequence by a web search for the Collatz conjecture.

Try some other starting values (e.g. $c_1 = 6$ or 13) and make a conjecture about the behaviour of this sequence for any starting value.

2 Using standard results

In the previous section you used two important results:

$$\sum_{r=1}^{n} 1 = n$$

$$\sum_{r=1}^{n} r = \tfrac{1}{2}n(n+1) \longleftarrow \boxed{\text{The sum of the integers.}}$$

There are similar results for the sum of the first n squares, and the first n cubes.

TECHNOLOGY

You could use a spreadsheet to verify these results for different values of n.

The sum of the squares: $\displaystyle\sum_{r=1}^{n} r^2 = \tfrac{1}{6}n(n+1)(2n+1)$

The sum of the cubes: $\displaystyle\sum_{r=1}^{n} r^3 = \tfrac{1}{4}n^2(n+1)^2$

These are important results. You will prove they are true later in the chapter.

These results can be used to sum other series, as shown in the following examples.

Example 4.3

(i) Write out the first three terms of the sequence $u_r = r^2 + 2r - 1$.

(ii) Find $\displaystyle\sum_{r=1}^{n} u_r$.

(iii) Use your answers from part (i) to check that your answer to part (ii) works for $n = 3$.

Solution

(i) $2, 7, 14$

(ii) $\displaystyle\sum_{r=1}^{n} u_r = \sum_{r=1}^{n}(r^2 + 2r - 1)$

$\displaystyle = \sum_{r=1}^{n} r^2 + 2\sum_{r=1}^{n} r - \sum_{r=1}^{n} 1$

$= \tfrac{1}{6}n(n+1)(2n+1) + 2 \times \tfrac{1}{2}n(n+1) - n$

$= \tfrac{1}{6}n[(n+1)(2n+1) + 6(n+1) - 6]$

$= \tfrac{1}{6}n(2n^2 + 3n + 1 + 6n + 6 - 6)$

$= \tfrac{1}{6}n(2n^2 + 9n + 1)$

> Look for common factors; here there is a common factor of n. You can also bring out the fraction $\tfrac{1}{6}$ by writing 1 as $\tfrac{6}{6}$. This means that the remaining expression is a quadratic with integer coefficients, which is much easier to factorise than a cubic with fractions in it.

(iii) $n = 3$

$\tfrac{1}{6}n(2n^2 + 9n + 1) = \tfrac{1}{6} \times 3 \times (18 + 27 + 1)$

$= \tfrac{1}{2} \times 46$

$= 23$

$2 + 7 + 14 = 23$

> It is a good idea to check your results like this, if you can.

Example 4.4

(i) Write the sum of this series using Σ notation.

$$(1 \times 3) + (2 \times 4) + (3 \times 5) + \ldots + n(n+2)$$

(ii) Hence find an expression for the sum in terms of n.

Solution

(i) $\displaystyle\sum_{r=1}^{n} r(r+2)$

(ii) $\displaystyle\sum_{r=1}^{n} r(r+2) = \sum_{r=1}^{n}(r^2 + 2r)$

$$= \sum_{r=1}^{n} r^2 + 2\sum_{r=1}^{n} r$$

$$= \tfrac{1}{6}n(n+1)(2n+1) + 2 \times \tfrac{1}{2}n(n+1)$$

$$= \tfrac{1}{6}n(n+1)[2n+1+6]$$

$$= \tfrac{1}{6}n(n+1)(2n+7)$$

Exercise 4.2

① (i) Write out the first three terms of the sequence $u_r = 2r - 1$.

   (ii) Find an expression for $\displaystyle\sum_{r=1}^{n}(2r - 1)$.

   (iii) Use part (i) to check part (ii).

② (i) Write out the first three terms of the sequence $u_r = r(3r + 1)$.

   (ii) Find an expression for $\displaystyle\sum_{r=1}^{n} r(3r + 1)$.

   (iii) Use part (i) to check part (ii).

③ (i) Write out the first three terms of the sequence $u_r = (r + 1)r^2$.

   (ii) Find an expression for $\displaystyle\sum_{r=1}^{n}(r + 1)r^2$.

   (iii) Use part (i) to check part (ii).

④ Find $\displaystyle\sum_{r=1}^{n}(4r^3 - 6r^2 + 4r - 1)$.

⑤ Find $(1 \times 2) + (2 \times 3) + (3 \times 4) + \ldots + n(n+1)$.

⑥ Find $(1 \times 2 \times 3) + (2 \times 3 \times 4) + (3 \times 4 \times 5) + \ldots + n(n+1)(n+2)$.

⑦ Find the sum of integers above n, up to and including $2n$, giving your answer in a fully factorised form.

⑧ Find the sum of the cubes of the integers larger than n, up to and including $3n$, giving your answer in a fully factorised form.

⑨ On a particularly artistic fruit stall, a pile of oranges is arranged to form a truncated square pyramid. Each layer is a square, with the lengths of the side of successive layers reducing by one orange (as in Figure 4.2).

The bottom layer measures $2n \times 2n$ oranges, and there are n layers.

(i) Prove that the number of oranges used is $\frac{1}{6}n(2n+1)(7n+1)$.

(ii) How many complete layers can the person setting up the stall use for this arrangement, given their stock of 1000 oranges? How many oranges are left over?

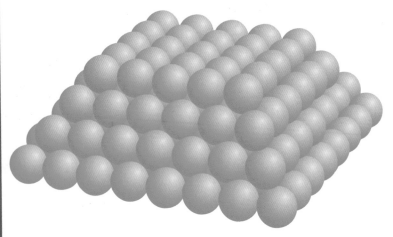

Figure 4.2

⑩ You have £20000 to invest for one year. You put it in the following bank account:

'Flexible Saver': 1.5% interest APR

- Interest calculated monthly (i.e. $\frac{1.5}{12}$% of balance each month).

- Interest paid annually, into a separate account.

- No limits on withdrawals or balance.

Your bank then informs you of a new savings account, which you are allowed to open as well as the Flexible Saver.

'Regular Saver': 5% interest APR

- Interest calculated monthly (i.e. $\frac{5}{12}$% each month).

- Interest paid annually, into a separate account.

- Maximum £1000 balance increase per month.

(i) Assuming you initially have your money in the Flexible Saver, but transfer as much as you can into a Regular Saver each month, calculate how much extra money you will earn, compared to what would happen if you just left it in the Flexible Saver all year.

(ii) Generalise your result – given an investment of I (in thousands of pounds), and a time of n months, what interest will you earn?

(Assume $n < I$, or you'll run out of funds to transfer.)

3 Mathematical induction

The oldest person to have ever lived, with documentary evidence, is believed to be a French woman called Jeanne Calment who died aged 122, in 1997.

Emily is an old woman who claims to have broken the record. A reporter asked her, 'How do you know you're 122 years old?'

She replied, 'Because I was 121 last year.'

Discussion point

➔ Is this a valid argument?

The sort of argument Emily was trying to use is called inductive reasoning. If all the elements are present it can be used in proof by induction. This is the subject of the rest of this chapter. It is a very beautiful form of proof but it is also very delicate; if you miss out any of the steps in the argument, as Emily did, you invalidate your whole proof.

ACTIVITY 4.1

Work out the first four terms of this pattern:

$$\frac{1}{1 \times 2} =$$

$$\frac{1}{1 \times 2} + \frac{1}{2 \times 3} =$$

$$\frac{1}{1 \times 2} + \frac{1}{2 \times 3} + \frac{1}{3 \times 4} =$$

$$\frac{1}{1 \times 2} + \frac{1}{2 \times 3} + \frac{1}{3 \times 4} + \frac{1}{4 \times 5} =$$

Activity 4.1 illustrates a common way of making progress in mathematics. Looking at a number of particular cases may show a pattern, which can be used to form a **conjecture** (i.e. a theory about a possible general result).

Conjectures are often written algebraically.

The conjecture can then be tested in further particular cases.

In this case, the sum of the first n terms of the sequence can be written as

$$\frac{1}{1 \times 2} + \frac{1}{2 \times 3} + \frac{1}{3 \times 4} + \ldots + \frac{1}{n(n + 1)}$$

The activity shows that the conjecture

$$\frac{1}{1 \times 2} + \frac{1}{2 \times 3} + \frac{1}{3 \times 4} + \ldots + \frac{1}{n(n + 1)} = \frac{n}{n + 1}$$

is true for $n = 1, 2, 3$ and 4.

Try some more terms, say, the next two.

If you find a **counter-example** at any point (a case where the conjecture is not true) then the conjecture is definitely disproved. If, on the other hand, the further cases agree with the conjecture then you may feel that you are on the right lines, but you can never be mathematically certain that trying another particular case might not reveal a counter-example: the conjecture is supported by more evidence but not proved.

The ultimate goal is to prove this conjecture is true for *all* positive integers. But it is not possible to prove this conjecture by deduction from known results. A different approach is needed; this is called **mathematical induction**.

In Activity 4.1 you established that the conjecture is true for some particular cases of n (specifically when $n = 1, 2, 3$ and 4) but this does not show that it is true for all positive integer values of n.

Now assume that the conjecture is true for a particular integer, $n = k$ say, so that

$$\frac{1}{1 \times 2} + \frac{1}{2 \times 3} + \frac{1}{3 \times 4} + \ldots + \frac{1}{k(k+1)} = \frac{k}{k+1}$$

and the idea is to use this assumption to check what happens for the next integer $n = k + 1$.

If the conjecture is true, then when $n = k + 1$ then you should get

$$\frac{1}{1 \times 2} + \frac{1}{2 \times 3} + \frac{1}{3 \times 4} + \ldots + \frac{1}{k(k+1)} + \frac{1}{(k+1)(k+2)} = \frac{(k+1)}{(k+1)+1} = \frac{k+1}{k+2}$$

> Replacing k by $k + 1$ in the result $\dfrac{k}{k+1}$

This is your target result. It is what you need to establish.

Look at the left-hand side (LHS). You can see that the first k terms are part of the assumption.

$$\frac{1}{1 \times 2} + \frac{1}{2 \times 3} + \frac{1}{3 \times 4} + \ldots + \frac{1}{k(k+1)} + \frac{1}{(k+1)(k+2)} \quad \text{(the LHS)}$$

$$= \frac{k}{k+1} + \frac{1}{(k+1)(k+2)} \qquad \boxed{\text{using the assumption}}$$

$$= \frac{k(k+2)+1}{(k+1)(k+2)} \qquad \boxed{\text{getting a common denominator}}$$

$$= \frac{k^2 + 2k + 1}{(k+1)(k+2)} \qquad \boxed{\text{expanding the top bracket}}$$

$$= \frac{(k+1)^2}{(k+1)(k+2)} \qquad \boxed{\text{factorising the top quadratic}}$$

$$= \frac{k+1}{k+2} \qquad \boxed{\text{cancelling the } (k+1) \text{ factor, since } k \neq -1}$$

$$\boxed{\text{which is the required result}}$$

These steps show that *if* the conjecture is true for $n = k$, *then* it is true for $n = k + 1$.

Since you have already proved it is true for $n = 1$, you can deduce that it is therefore true for $n = 2$ (by taking $k = 1$).

You can continue in this way (e.g. take $k = 2$ and deduce that it is true for $n = 3$), as far as you want to go. Since you can reach *any* positive integer n, you have now proved that the conjecture is true for *every* positive integer.

This method of **proof by mathematical induction** (often shortened to **proof by induction**) is a bit like the process of climbing a ladder:

If you can:

1 get on the ladder (the bottom rung)
2 get from one rung to the next

then you can climb as far up the ladder as you like.

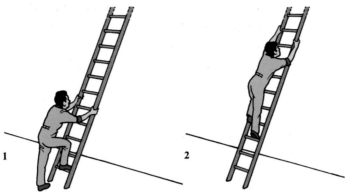

Figure 4.3

The corresponding steps in the previous proof are:

1 showing that the conjecture is true for $n = 1$ and
2 showing that *if* it is true for a particular value ($n = k$, say), *then* it is true for the next one ($n = k + 1$).

(Notice the *if ... then ...* structure to this step).

You should conclude any argument by mathematical induction with a statement of what you have shown.

Steps in proof by induction

To prove something by mathematical induction you need to state a conjecture to begin with. There are then five elements needed to try to prove that the conjecture is true.

- Proving that it is true for a starting value (usually $n = 1$).

- Assuming the result is true for $n = k$ and finding the target expression: using the result for $n = k$ to find the equivalent expression for $n = k + 1$.

> This can be done before or after finding the target expression, but you may find it easier to find the target expression first, so that you know what you are working towards

> To find the target expression you replace k with $k + 1$ in the result for $n = k$

- Proving that:
  if it is true for $n = k$, *then* it is true for $n = k + 1$.

- Arguing that since it is true for $n = 1$, it is also true for $n = 1 + 1 = 2$, and then for $n = 2 + 1 = 3$, and all subsequent values of n.

- Concluding the argument by writing down the result and stating that it has been proved, by induction.

> This ensures the argument is properly rounded off. You will often use the word 'therefore'

The sum of the squares of the first n integers

Example 4.5

Prove that, for all positive integers n:

$$1^2 + 2^2 + 3^2 + \ldots + n^2 = \frac{1}{6}n(n + 1)(2n + 1)$$

Solution

When $n = 1$, $\quad$ LHS $= 1^2 = 1$ $\quad$ RHS $= \frac{1}{6} \times 1 \times 2 \times 3 = 1$

So it is true for $n = 1$.

Assume the result is true for $n = k$,

so $\quad 1^2 + 2^2 + 3^2 + \ldots + k^2 = \frac{1}{6}k(k + 1)(2k + 1)$

Target expression:

$$1^2 + 2^2 + 3^2 + \ldots + k^2 + (k+1)^2 = \tfrac{1}{6}(k+1)\big((k+1)+1\big)\big(2(k+1)+1\big)$$

$$= \tfrac{1}{6}(k+1)(k+2)(2k+3)$$

You want to prove that *if* the result is true for $n = k$ then it is true for $n = k + 1$.

Look at the LHS of the result you want to prove:

$$1^2 + 2^2 + 3^2 + \ldots + k^2 + (k+1)^2$$

Use the assumed result for $n = k$ to replace the first k terms:

$$= \tfrac{1}{6}k(k+1)(2k+1) + (k+1)^2 \quad \longleftarrow \boxed{\text{The } (k+1)\text{th term}}$$

$$= \tfrac{1}{6}(k+1)[k(2k+1) + 6(k+1)] \quad \boxed{\text{First } k \text{ terms}}$$

$$= \tfrac{1}{6}(k+1)\big[2k^2 + 7k + 6\big]$$

$$= \tfrac{1}{6}(k+1)(k+2)(2k+3) \quad \longleftarrow \boxed{\begin{array}{l}\text{This is the same as the target}\\ \text{expression, as required.}\end{array}}$$

> Take out common factor $\tfrac{1}{6}(k+1)$. You can see from the target expression that this is useful.

If the result is true for $n = k$ then it is true for $n = k + 1$.

Since true for $n = 1$ and if true for $n = k$ then true for $n = k + 1$, so it is true for all positive integer values.

Therefore the result that $1^2 + 2^2 + 3^2 + \ldots + n^2 = \tfrac{1}{6}n(n+1)(2n+1)$ is true for all $n \in \mathbb{Z}^+$.

ACTIVITY 4.2

Jane is investigating the sum of the first n even numbers.

She writes: $2 + 4 + 6 + \ldots + 2n = \left(n + \dfrac{1}{2}\right)^2$.

(i) Prove that *if* this result is true when $n = k$, *then* it is true when $n = k + 1$. Explain why Jane's conjecture is *not* true for all positive integers n.

(ii) Suggest a different conjecture for the sum of the first n even numbers, that is true for $n = 1$ but not for other values of n. At what point does an attempt to use proof by induction on this result break down?

Exercise 4.3

① (i) Show that the result $1 + 3 + 5 + \ldots + (2n - 1) = n^2$ is true for the case $n = 1$.

(ii) Assume that $1 + 3 + 5 + \ldots + (2k - 1) = k^2$ and use this to prove that $1 + 3 + 5 + \ldots + (2k - 1) + (2k + 1) = (k + 1)^2$.

(iii) Explain how parts (i) and (ii) together prove that the sum of the first n odd integers is n^2.

② (i) Show that the result $1 + 5 + 9 + \dots + (4n - 3) = n(2n - 1)$ is true for the case $n = 1$.

(ii) Assume that $1 + 5 + 9 + \dots + (4k - 3) = k(2k - 1)$ and use this to prove that $1 + 5 + 9 + \dots + (4k - 3) + (4(k + 1) - 3) = (k + 1)(2(k + 1) - 1)$.

(iii) Explain how parts (i) and (ii) together prove that $1 + 5 + 9 + \dots + (4n - 3) = n(2n - 1)$ for all positive integers n.

Prove the following results, for $n \in \mathbb{Z}^+$, by induction.

③ $1 + 2 + 3 + \dots + n = \frac{1}{2}n(n + 1)$ (the sum of the first n integers)

④ $\displaystyle\sum_{r=1}^{n} r^3 = \frac{1}{4}n^2(n + 1)^2$ (the sum of the first n cubes)

⑤ $2^1 + 2^2 + 2^3 + 2^4 + \dots + 2^n = 2(2^n - 1)$

⑥ $\displaystyle\sum_{r=0}^{n} x^r = \frac{1 - x^{n+1}}{1 - x}$ $(x \neq 1)$

⑦ $(1 \times 2 \times 3) + (2 \times 3 \times 4) + \dots + n(n + 1)(n + 2) = \frac{1}{4}n(n + 1)(n + 2)(n + 3)$

⑧ $\displaystyle\sum_{r=1}^{n} (3r + 1) = \frac{1}{2}n(3n + 5)$

Prove the following results by induction.

⑨ $\frac{1}{3} + \frac{1}{15} + \frac{1}{35} + \dots + \frac{1}{4n^2 - 1} = \frac{n}{2n + 1}$ for $n \in \mathbb{Z}^+$

⑩ $\left(1 - \frac{1}{2^2}\right)\left(1 - \frac{1}{3^2}\right)\left(1 - \frac{1}{4^2}\right) \cdots \left(1 - \frac{1}{n^2}\right) = \frac{n + 1}{2n}$ for $n \in \mathbb{Z}^+, n \geqslant 2$

⑪ $1 \times 1! + 2 \times 2! + 3 \times 3! + \dots + n \times n! = (n + 1)! - 1$

⑫ (i) Prove by induction that $\displaystyle\sum_{r=1}^{n} (5r^4 + r^2) = \frac{1}{2}n^2(n + 1)^2(2n + 1)$ for $n \in \mathbb{Z}^+$.

(ii) Using the result in part (i) and the formula $\displaystyle\sum_{r=1}^{n} r^2 = \frac{1}{6}n(n + 1)(2n + 1)$,

show that $\displaystyle\sum_{r=1}^{n} r^4 = \frac{1}{30}n(n + 1)(2n + 1)(3n^2 + 3n - 1)$ for $n \in \mathbb{Z}^+$.

4 Other proofs by induction

So far you have used induction to prove a given expression for the sum of a series. Here are some other examples of its use.

In Example 4.6 a sequence is given using the term-to-term form and the position-to-term form has to be proved. This idea is developed further in A level Further Mathematics.

Example 4.6

A sequence is defined by $u_{n+1} = 4u_n - 3$, $u_1 = 2$.

Prove that $u_n = 4^{n-1} + 1$.

Solution

For $n = 1$: $u_1 = 4^0 + 1 = 1 + 1 = 2$, so the result is true for $n = 1$.

Assume that the result is true for $n = k$, so that $u_k = 4^{k-1} + 1$.

Target expression:
$u_{k+1} = 4^k + 1$

For $n = k + 1$:
$$\begin{aligned} u_{k+1} &= 4u_k - 3 \\ &= 4\left(4^{k-1} + 1\right) - 3 \\ &= 4 \times 4^{k-1} + 4 - 3 \\ &= 4^k + 1 \end{aligned}$$

If the result is true for $n = k$ then it is true for $n = k + 1$.

Since true for $n = 1$ and if true for $n = k$ then true for $n = k + 1$, so it is true for all positive integer values.

Therefore the result that $u_n = 4^{n-1} + 1$ is true for all $n \in \mathbb{Z}^+$.

You can use induction to prove results about powers of certain matrices.

Example 4.7

Given $\mathbf{A} = \begin{pmatrix} 4 & 1 \\ 3 & 2 \end{pmatrix}$, prove by induction that

$$\mathbf{A}^n = \frac{1}{4}\begin{pmatrix} 3 \times 5^n + 1 & 5^n - 1 \\ 3 \times 5^n - 3 & 5^n + 3 \end{pmatrix}.$$

Solution

For $n = 1$: $\text{LHS} = \mathbf{A}^1 = \begin{pmatrix} 4 & 1 \\ 3 & 2 \end{pmatrix}$

$$\text{RHS} = \frac{1}{4}\begin{pmatrix} 3 \times 5 + 1 & 5 - 1 \\ 3 \times 5 - 3 & 5 + 3 \end{pmatrix} = \frac{1}{4}\begin{pmatrix} 16 & 4 \\ 12 & 8 \end{pmatrix}$$

$$= \begin{pmatrix} 4 & 1 \\ 3 & 2 \end{pmatrix} \quad \text{so the result is true for } n = 1.$$

Assume that the result is true for $n = k$, so that

$$\mathbf{A}^k = \frac{1}{4}\begin{pmatrix} 3 \times 5^k + 1 & 5^k - 1 \\ 3 \times 5^k - 3 & 5^k + 3 \end{pmatrix}$$

Target expression:
$$\mathbf{A}^{k+1} = \frac{1}{4}\begin{pmatrix} 3 \times 5^{k+1} + 1 & 5^{k+1} - 1 \\ 3 \times 5^{k+1} - 3 & 5^{k+1} + 3 \end{pmatrix}$$

For $n = k + 1$:

$$\mathbf{A}^{k+1} = \mathbf{A}^k \mathbf{A}$$

$$= \frac{1}{4}\begin{pmatrix} 3 \times 5^k + 1 & 5^k - 1 \\ 3 \times 5^k - 3 & 5^k + 3 \end{pmatrix}\begin{pmatrix} 4 & 1 \\ 3 & 2 \end{pmatrix}$$

$$= \frac{1}{4}\begin{pmatrix} 12 \times 5^k + 4 + 3 \times 5^k - 3 & 3 \times 5^k + 1 + 2 \times 5^k - 2 \\ 12 \times 5^k - 12 + 3 \times 5^k + 9 & 3 \times 5^k - 3 + 2 \times 5^k + 6 \end{pmatrix}$$

Multiplying matrices

$$= \frac{1}{4}\begin{pmatrix} 15 \times 5^k + 1 & 5 \times 5^k - 1 \\ 15 \times 5^k - 3 & 5 \times 5^k + 3 \end{pmatrix}$$

Using $15 = 3 \times 5$

$$= \frac{1}{4}\begin{pmatrix} 3 \times 5^{k+1} + 1 & 5^{k+1} - 1 \\ 3 \times 5^{k+1} - 3 & 5^{k+1} + 3 \end{pmatrix}$$

This is the target matrix

as required

If the result is true for $n = k$ then it is true for $n = k + 1$.

Since true for $n = 1$ and if true for $n = k$ then true for $n = k + 1$, so it is true for all positive integer values.

Therefore the result that $\mathbf{A}^n = \frac{1}{4}\begin{pmatrix} 3 \times 5^n + 1 & 5^n - 1 \\ 3 \times 5^n - 3 & 5^n + 3 \end{pmatrix}$ is true for all $n \in \mathbb{Z}^+$.

Another use of proof by induction is to check divisibility, as the following example shows.

Example 4.8

Prove that $4^n + 6n - 1$ is divisible by 9 for all $n \in \mathbb{Z}^+$.

Solution

For $n = 1$:　$4^1 + 6(1) - 1 = 4 + 6 - 1 = 9$　so the result is true for $n = 1$.

Assume that the result is true for $n = k$, so that $4^k + 6k - 1 = 9N$ for some integer N.

For $n = k + 1$:

$$4^{k+1} + 6(k + 1) - 1 = \left\{4\left(4^k + 6k - 1\right) - 24k + 4\right\} + 6(k + 1) - 1$$

Writing in terms of the assumed result

$$= 4 \times 9N - 24k + 4 + 6k + 6 - 1$$

$$= 9 \times 4N - 18k + 9 = 9(4N - 2k + 1)$$

This is a multiple of 9, as required

If the result is true for $n = k$ then it is true for $n = k + 1$.

Since true for $n = 1$ and if true for $n = k$ then true for $n = k + 1$, so it is true for all positive integer values.

Therefore $4^n + 6n - 1$ is divisible by 9 for all $n \in \mathbb{Z}^+$.

Exercise 4.4

① A sequence is defined by $u_{n+1} = 3u_n + 2$, $u_1 = 2$. Prove by induction that $u_n = 3^n - 1$.

② Given that $M = \begin{pmatrix} 2 & 0 \\ 0 & 3 \end{pmatrix}$, prove by induction that $M^n = \begin{pmatrix} 2^n & 0 \\ 0 & 3^n \end{pmatrix}$, for $n \in \mathbb{Z}^+$.

③ $f(n) = 2^n + 6^n$.
  (i) Show that $f(k + 1) = 6f(k) - 4(2^k)$.
  (ii) Hence, or otherwise, prove by induction that, for $n \in \mathbb{Z}^+$, $f(n)$ is divisible by 4.

④ A sequence is defined by $u_{n+1} = \dfrac{u_n}{u_n + 1}$, $u_1 = 1$.
  (i) Find the values of u_2, u_3 and u_4.
  (ii) Suggest a general formula for u_n and prove your conjecture by induction.

⑤ You are given the matrix $\mathbf{A} = \begin{pmatrix} -1 & -4 \\ 1 & 3 \end{pmatrix}$.
  (i) Calculate $\mathbf{A}^2$ and $\mathbf{A}^3$.
  (ii) Show that the formula $\mathbf{A}^n = \begin{pmatrix} 1 - 2n & -4n \\ n & 1 + 2n \end{pmatrix}$ is consistent with the given value of $\mathbf{A}$ and your expressions for $\mathbf{A}^2$ and $\mathbf{A}^3$.
  (iii) Prove by induction that the formula for $\mathbf{A}^n$ is correct for all positive integers n.

⑥ Prove by induction that $2^{4n+1} + 3$ is a multiple of 5 for all positive integers n.

⑦ (i) Prove by induction that, for $n \in \mathbb{Z}^+$, $\displaystyle\sum_{r=1}^{n} r(r + 2) = \frac{1}{6}n(n + 1)(2n + 7)$.
  (ii) A sequence of positive integers is defined by $u_1 = 2$, $u_{n+1} = u_n + 2(n + 1)$, for $n \in \mathbb{Z}^+$. Prove by induction that $u_n = n(n + 1)$, $n \in \mathbb{Z}^+$.

⑧ Prove that $11^{n+2} + 12^{2n+1}$ is divisible by 133 for all integers $n \geqslant 0$.

⑨ You are given the matrix $\mathbf{M} = \begin{pmatrix} -1 & 2 \\ 3 & 1 \end{pmatrix}$.
  (i) Calculate $\mathbf{M}^2$, $\mathbf{M}^3$ and $\mathbf{M}^4$.
  (ii) Write down separate conjectures for formulae for $\mathbf{M}^n$ for even n (i.e. $\mathbf{M}^{2m}$) and for odd n (i.e. $\mathbf{M}^{2m+1}$).
  (iii) Prove each conjecture by induction on the value of m.

⑩ Let $F_n = 2^{(2^n)} + 1$
  (i) Calculate F_0, F_1, F_2, F_3 and F_4.
  (ii) Prove by induction that $F_0 \times F_1 \times F_2 \times \ldots \times F_{n-1} = F_n - 2$.
  (iii) Use the result from part (ii) to prove that F_i and F_j are coprime (have no common factors other than 1) for all i, j $(i \neq j)$.
  (iv) Use the result from part (iii) to prove that there are infinitely many prime numbers. The F_n numbers are called Fermat numbers. The first five are prime (the Fermat primes). Nobody (yet) knows if any of the other Fermat numbers are prime.

LEARNING OUTCOMES

When you have completed this chapter you should be able to:

➤ know what is meant by a sequence and a series

➤ find the sum of a series using standard formulae for $\sum r$, $\sum r^2$ and $\sum r^3$

➤ find the sum of a series using the method of differences

➤ use proof by induction to prove given results for the sum of a series

➤ use proof by induction to prove given results for the nth term of a sequence

➤ use proof by induction to prove given results for the nth power of a matrix.

KEY POINTS

1 The terms of a sequence are often written as a_1, a_2, a_3, ... or u_1, u_2, u_3, ... The general term of a sequence may be written as a_r or u_r [sometimes the letters k or i are used instead of r]. The last term is usually written as a_n or u_n.

2 A series is the sum of the terms of a sequence. The sum S_n of the first n terms of a sequence can be written as $S_n = a_1 + a_2 + a_3 + \ldots + a_n = \sum_{r=1}^{n} a_r$.

3 Some series can be expressed as combinations of these standard results:

$$\sum_{r=1}^{n} r = \frac{1}{2}n(n+1) \quad \sum_{r=1}^{n} r^2 = \frac{1}{6}n(n+1)(2n+1) \quad \sum_{r=1}^{n} r^3 = \frac{1}{4}n^2(n+1)^2$$

4 To prove, by induction, that a statement involving an integer n is true for all $n \geqslant n_0$, you need to:

■ prove that the result is true for an initial value $n = n_0$ (usually $n = 1$)

■ assume that the result is true for some general value $n = k$ and find the target expression (the result when $n = k + 1$)

■ prove (algebraically) that if the result is true for $n = k$ then it is true for $n = k + 1$

■ argue that since the result is true for $n = n_0$ and *if* the result is true for $n = k$ *then* it is true for $n = k + 1$, it must be true for $n = n_0 + 1$, $n = n_0 + 2$, ...

■ conclude the argument with a precise statement about what has been proved.

FUTURE USES

■ You will develop the work on sequences and recursion in A level Further Mathematics.

Practice Questions 1

For questions 1 to 5 you must show non-calculator methods in your answer.

① (i) Calculate the matrix product

$$\begin{pmatrix} 1 & 3 & 2 \\ 2 & 0 & -1 \end{pmatrix} \begin{pmatrix} 1 & -1 \\ 2 & 0 \\ 0 & 3 \end{pmatrix}$$

[3 marks]

(ii) $\mathbf{A} = \begin{pmatrix} 1 & 2 \\ 3 & 1 \end{pmatrix}$

Calculate $\mathbf{A}^2 + 4\mathbf{A}$ [3 marks]

② (i) The complex number $w = 1 + 2i$. Plot, on a single Argand diagram, the points which represent the four complex numbers: $w, w^2, w - w^*$ and $\dfrac{1}{w} + \dfrac{1}{w^*}$. [5 marks]

(ii) Which two of the numbers: $w, w^2, w - w^*$ and $\dfrac{1}{w} + \dfrac{1}{w^*}$ have the same imaginary part? [1 mark]

③ One of the roots of the cubic equation $z^3 - 9z^2 + 28z - 30 = 0$ is $3 + i$. Another root of the cubic equation is an integer. Solve the cubic equation. [5 marks]

④ Ezra is investigating whether the formula for solving quadratic equations works if the coefficients of the quadratic are not real numbers.

Here is the beginning of his working for the quadratic equation $(2 + i)z^2 + 6z + (2 - i) = 0$.

$$z = \frac{-b \pm \sqrt{b^2 - 4ac}}{2a}$$

$$= \frac{-6 \pm \sqrt{36 - 4(2 + i)(2 - i)}}{2(2 + i)}$$

$$= \dots$$

(i) Finish Ezra's working. Show that both of the answers given by this method are of the form $\lambda(2 - i)$, where λ is real, stating the value of λ in each case. [4 marks]

(ii) How should Ezra check that his answers are indeed roots of the equation? [1 mark]

⑤ The cubic equation $x^3 + 3x^2 - 6x - 8 = 0$ has roots α, β, γ.

(i) Find a cubic equation with roots $\alpha + 1, \beta + 1, \gamma + 1$. [4 marks]

(ii) Solve the equation you found as your answer to part (i). [3 marks]

(iii) Solve the equation $x^3 + 3x^2 - 6x - 8 = 0$. [2 marks]

PS ⑥ (i) What transformation is represented by the matrix $\mathbf{B} = \begin{pmatrix} 0 & -1 \\ 1 & 0 \end{pmatrix}$? [2 marks]

(ii) By considering transformations, or otherwise, find a matrix $\mathbf{A}$ such that $\mathbf{A}^2 = \mathbf{B}$. [3 marks]

MP PS ⑦ The quadratic equation $az^2 + bz + c = 0$ has roots δ and $\delta + 1$.
By considering the sum and product of its roots, or otherwise,
prove that $b^2 - 4ac = a^2$. [5 marks]

MP ⑧ A sequence is defined by $u_{k+1} = 2u_k - k + 1$ with $u_1 = 3$.

(i) Write down the first five terms of the sequence. [1 mark]

(ii) Prove by induction that $u_n = 2^n + n$. [5 marks]

⑨ The matrix $\mathbf{R} = \begin{pmatrix} -\dfrac{3}{5} & \dfrac{4}{5} \\ \dfrac{4}{5} & \dfrac{3}{5} \end{pmatrix}$ represents a transformation R.

The unit square OIPJ has coordinates $O(0, 0), I(1, 0), P(1, 1), J(0, 1)$.

(i) Plot, on the same diagram, the unit square and its image
$O'I'P'J'$ under R. [2 marks]

(ii) Find the equation of the line of invariant points for R. [3 marks]

(iii) Verify that the line which is perpendicular to this line
of invariant points, and which passes through the origin,
is an invariant line. [3 marks]

(iv) Mark on your diagram two points on the unit square
which are invariant under R. [2 marks]

MP ⑩ (i) Prove by induction that $f(n) = \dfrac{n}{3} + \dfrac{n^2}{2} + \dfrac{n^3}{6}$ is integer-valued
for all positive integer values of n. [5 marks]

(ii) Write $\dfrac{n}{3} + \dfrac{n^2}{2} + \dfrac{n^3}{6}$ as a single fraction and explain how
this shows that it $f(n)$ is integer-valued for all positive
integer values of n. [3 marks]

MP ⑪ (i) $\mathbf{M} = \begin{pmatrix} 2 & -1 \\ 0 & 1 \end{pmatrix}$. Prove by induction that

$\mathbf{M}^n = \begin{pmatrix} 2^n & 1 - 2^n \\ 0 & 1 \end{pmatrix}$. [5 marks]

(ii) Prove by induction that $5^n - 4n - 1$ is divisible by 8. [5 marks]

5 Applications of integration

Discussion point

→ What plane shape would need to be rotated through 360° to produce a solid in the shape of an egg?

1 Volumes of revolution

When the shaded region in Figure 5.1 is rotated by 360° about the x-axis the solid obtained, illustrated in Figure 5.2, is called a solid of revolution.

Discussion points

→ How could you use the formula for the volume of a cone to work out the volume of this solid?

→ Do you know formulae that will allow you to find the volumes of any other solids of revolution?

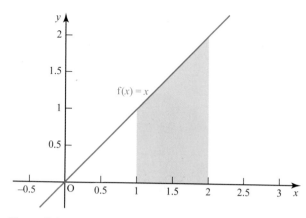

Figure 5.1

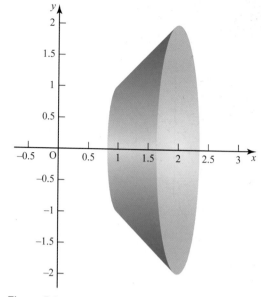

Figure 5.2

If the line $y = x$ in Figure 5.1 is replaced with a curve then the calculation is less obvious.

You already know that you can use integration to calculate the area under a curve. In Figure 5.3, each of the rectangles has a height of y (which depends on the value of x) and a width of δx, where δx is small.

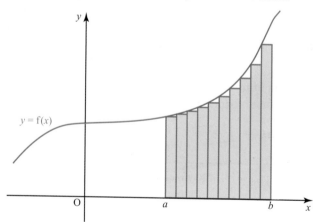

Figure 5.3

The total area of the rectangles is given by $\sum_{x=a}^{x=b} y\, \delta x$.

As the rectangles become thinner, the approximation for the area becomes more accurate. In the limiting case, as $\delta x \to 0$, the sum becomes an integral and the expression for the area is exact.

$$A = \int_a^b y\, dx.$$

The same ideas can be used to find the volume of a solid of revolution.

Look at the shaded region in Figure 5.4, and the solid of revolution it would form when rotated 360° about the x-axis in Figure 5.5. When the green strip is rotated, it forms a disc.

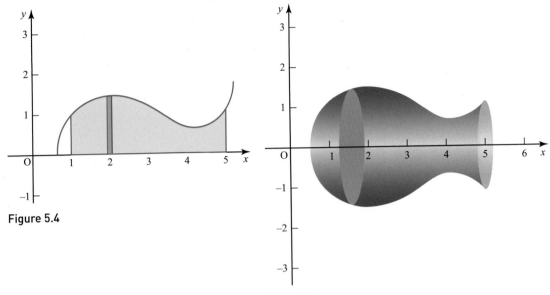

Figure 5.4

Figure 5.5

The green disc shown in Figure 5.5 is approximately cylindrical, with radius y and thickness δx so its volume is given by

$$\delta V = \pi y^2 \delta x$$

The volume of the complete solid is then approximately the sum of all these discs, and this approximation will be better as the thickness of the discs gets smaller.

$$V \approx \sum \delta V$$

$$V \approx \sum_{x=a}^{x=b} \pi y^2 \delta x$$

The limit of this sum, as $\delta x \to 0$, becomes an integral, and you then have an accurate formula for the volume of revolution:

$$V = \int_a^b \pi y^2 \, \mathrm{d}x$$

Example 5.1

The region between the curve $y = x^2$, the x-axis and the lines $x = 1$ and $x = 3$ is rotated through $360°$ about the x-axis.

Find the volume of revolution which is formed.

Solution

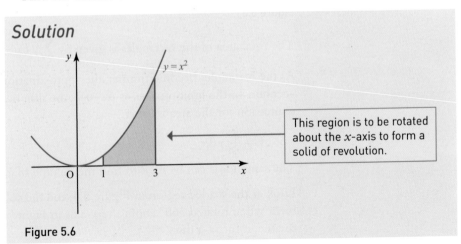

This region is to be rotated about the x-axis to form a solid of revolution.

Figure 5.6

Using the formula $V = \int_a^b \pi y^2 \, dx$

$$V = \int_1^3 \pi (x^2)^2 \, dx$$

Replacing y with x^2.

$$= \pi \int_1^3 x^4 \, dx$$

$$= \pi \left[\frac{x^5}{5} \right]_1^3$$

$$= \frac{\pi}{5} (243 - 1)$$

Depending on the circumstances, you may want to write this as 152 cubic units (to 3 s.f.). Unless a decimal answer is required, it is common to leave π in the answer, and so keep the answer exact.

$$= \frac{242\pi}{5}$$

Rotation around the y-axis

You can also form a solid of revolution by rotating a region about the y-axis. The diagram in Figure 5.8 shows the solid which is obtained when a region between part of the curve from Figure 5.7 and the y-axis is rotated through 360° about the y-axis.

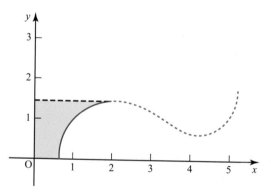

Figure 5.7

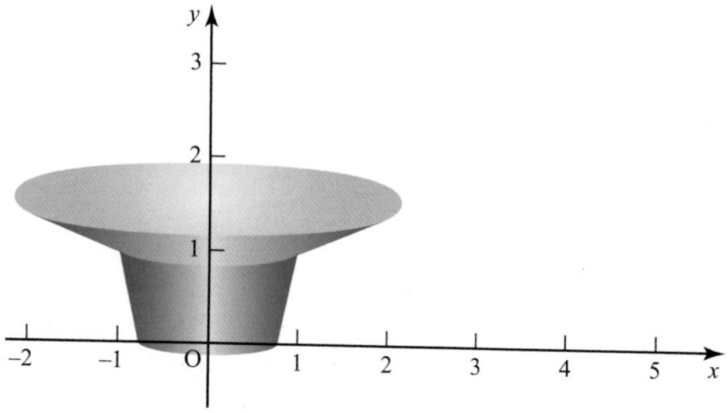

Figure 5.8

The formula for the volume can be obtained in a similar way, so that

$$V_{y\text{-}axis} = \int \pi x^2 \, dy$$

The limits in this case are y values rather than x-values.

$$V_{y\text{-}axis} = \int_{y=q}^{y=p} \pi x^2 \, dy$$

Since the integration is with respect to y, the limits can just be written as p and q.

$$V_{y\text{-}axis} = \int_q^p \pi x^2 \, dy$$

You would need to write x in terms of y so that you can integrate with respect to y.

| Example 5.2 | The region between the curve $y = x^2$, the y-axis and the lines $y = 2$ and $y = 5$ is rotated 360° around the y-axis. Find the volume of revolution obtained. |

Solution

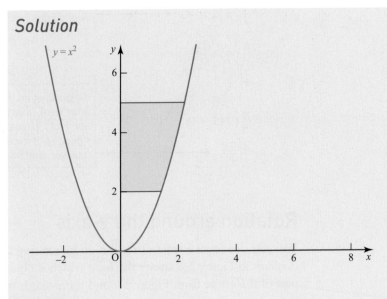

Figure 5.9

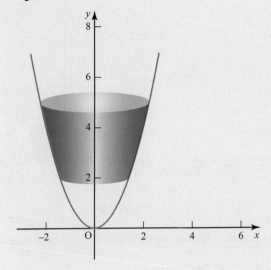

Figure 5.10

Using $V = \displaystyle\int_q^p \pi x^2 \, dy$

$V = \displaystyle\int_2^5 \pi y \, dy$ ← $\boxed{\text{Notice that } x^2 = y.}$

$\qquad = \pi \left[\dfrac{y^2}{2} \right]_2^5$

$\qquad = \pi \left(\dfrac{25}{2} - \dfrac{4}{2} \right)$

$\qquad = \dfrac{21\pi}{2}$ cubic units

Exercise 5.1

① Name six common objects which could be generated as solids of revolution.

② Figure 5.11 shows the line $y = 3x$.

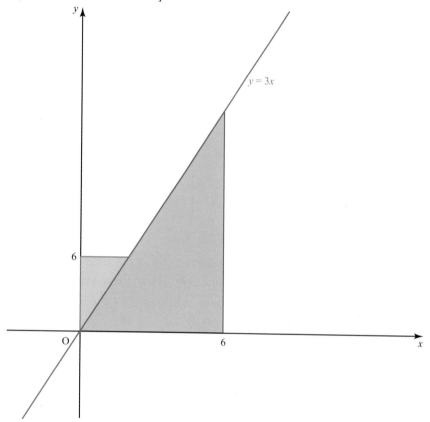

Figure 5.11

(i) Describe the solid obtained by rotating the blue region by 360° around the x-axis.

(ii) Use the formula $V = \pi \int_{x_1}^{x_2} y^2 \, dx$ to calculate the volume of the solid.

(iii) Describe the solid obtained by rotating the orange region by 360° around the y-axis.

(iv) Use the formula $V = \pi \int_{y_1}^{y_2} x^2 \, dy$ to calculate the volume of the solid.

(v) Use the formula for the volume of a cone to show that both your answers are correct.

③ Figure 5.12 shows the region under the curve $y = \dfrac{1}{x}$ between $x = 1$ and $x = 2$.

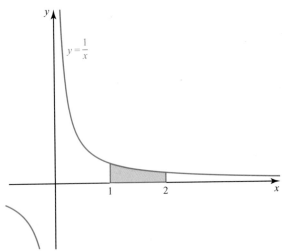

Figure 5.12

Find the volume of the solid of revolution formed by rotating this region through 360° about the x-axis.

④ A region is bounded by the lines $y = x + 2$, the x-axis and the y-axis and the line $x = 2$.

   (i)   Draw a sketch to show this region.

   (ii)   Find the volume of the solid obtained by rotating the region through 360° about the x-axis.

⑤ A region is bounded by the curve $y = x^2 - 1$ and the x-axis.

   (i)   Draw a sketch to show this region.

   (ii)   Find the volume of the solid obtained by rotating the region through 360° about the x-axis.

⑥ A region is bounded by the curve $y = \sqrt{x}$, the y-axis and the line $y = 2$.

   (i)   Draw a sketch to show this region.

   (ii)   Find the volume of the solid obtained by rotating the region through 360° about the y-axis.

⑦ (i)   Sketch the graph of $y = (x - 3)^2$ for values of x between $x = -1$ and $x = 5$.

        Shade in the region under your curve between $x = 0$ and $x = 2$.

   (ii)   Calculate the area of the shaded region.

   (iii)   The shaded region is rotated about the x-axis to form a volume of revolution. Calculate this volume.

⑧ A mathematical model for a large plant pot is obtained by rotating the part of the curve $y = 0.1x^2$ which is between $x = 10$ and $x = 25$ through 360° about the y-axis and then adding a flat base. Units are in centimetres.

   (i)   Draw a sketch of the curve and shade in the area of the cross-section of the pot, indicating which line will form its base.

   (ii)   Garden compost is sold in litres. How many litres will be required to fill the pot to a depth of 45 cm? (Ignore the thickness of the pot.)

⑨ Figure 5.13 shows the curve $y = x^2 - 4$.

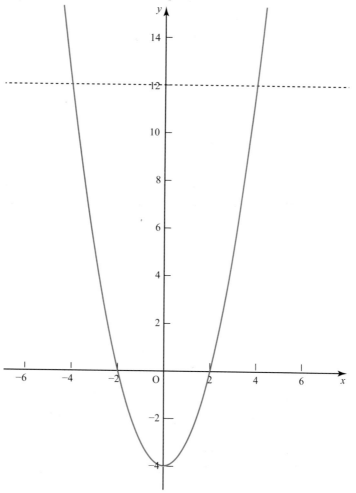

Figure 5.13

The region R is formed by the line $y = 12$, the y-axis, the x-axis, and the curve $y = x^2 - 4$, for positive values of x.

(i) Make a sketch copy of the graph and shade the region R.

The inside of a vase is formed by rotating the region R through $360°$ about the y-axis. Each unit of x and y represents $2\,\text{cm}$.

(ii) Write down an expression for the volume of revolution of the region R about the y-axis.

(iii) Find the capacity of the vase in litres.

(iv) Show that when the vase is filled to $\dfrac{5}{6}$ of its internal height it is three-quarter full.

⑩ Figure 5.14 shows the circle $x^2 + y^2 = 25$ and the line $y = 4$.

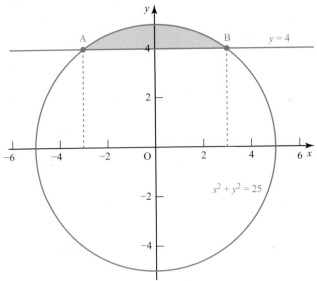

Figure 5.14

(i) Find the coordinates of the points A and B where the circle and line intersect.

(ii) A napkin ring is formed by rotating the shaded area through 360° about the x-axis. Find the volume of the napkin ring.

⑪ The function f(x) describes a semicircle, radius r, at a distance R above the origin:

$$f(x) = \sqrt{r^2 - x^2} + R$$

Use this function, and a similar function describing the other half of the circle, and your knowledge of volumes of revolution, to prove that the volume of a torus, as shown in Figure 5.15, is $2\pi^2 r^2 R$ or equivalently, $(2\pi R) \times (\pi r^2)$.

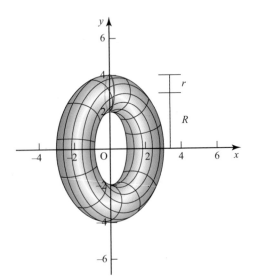

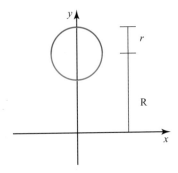

Figure 5.16

Figure 5.15

LEARNING OUTCOMES

When you have completed this chapter you should be able to:

➤ derive formulae for and calculate the volumes of solids generated by rotating a plane region about the x-axis

➤ derive formulae for and calculate the volumes of solids generated by rotating a plane region about the y-axis

KEY POINTS

1 The volume of revolution when a curve is rotated through 2π about the x-axis, between $x = a$ and $x = b$ is $V_x = \pi \displaystyle\int_a^b y^2 \, dx$.

2 The volume of revolution when a curve is rotated through 2π about the y-axis, between $y = p$ and $y = q$ is $V_y = \pi \displaystyle\int_p^q x^2 \, dy$.

3 If the curve is defined in terms of a parameter t, change the limits to t-values and replace dx by $\dfrac{dx}{dt} dt$ or dy by $\dfrac{dy}{dt} dt$.

FUTURE USES

■ In A2 Further Mathematics you will calculate the surface area of a volume of revolution.

Complex numbers and geometry

The power of mathematics is often to change one thing into another, to change geometry into language.

Marcus du Sautoy

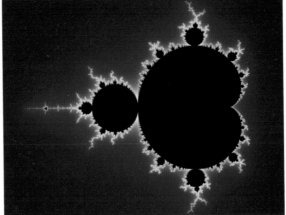

Figure 6.1 The Mandlebrot set

Discussion point

→ Figure 6.1 is an Argand diagram showing the Mandlebrot set. The black area shows all the complex numbers that satisfy a particular rule. Find out about the rule which defines whether or not a particular complex number is in the Mandlebrot set.

1 The modulus and argument of a complex number

Figure 6.2 shows the point representing $z = x + y$i on an Argand diagram.

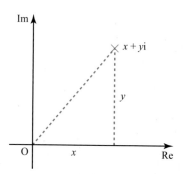

Figure 6.2

> Using Pythagoras' theorem.

The distance of this point from the origin is $\sqrt{x^2 + y^2}$.

This distance is called the modulus of z, and is denoted by $|z|$.

So, for the complex number $z = x + y$i, $|z| = \sqrt{x^2 + y^2}$.

Notice that since $zz^* = (x + i y)(x - i y) = x^2 + y^2$, then $|z|^2 = zz^*$.

Example 6.1

Represent each of the following complex numbers on an Argand diagram. Find the modulus of each complex number, giving exact answers in their simplest form.

$$z_1 = -5 + i \qquad z_2 = 6 \qquad z_3 = -5 - 5i \qquad z_4 = -4i$$

Solution

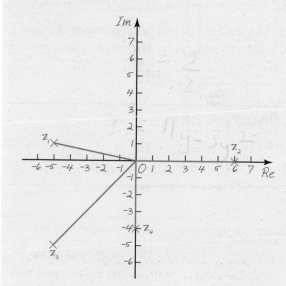

Figure 6.3

$$|z_1| = \sqrt{(-5)^2 + 1} = \sqrt{26}$$

$$|z_2| = \sqrt{6^2 + 0^2} = \sqrt{36} = 6$$

$$|z_3| = \sqrt{(-5)^2 + (-5)^2} = \sqrt{50} = 5\sqrt{2}$$

$$|z_4| = \sqrt{0^2 + (-4)^2} = \sqrt{16} = 4$$

Notice that the modulus of a real number $z = a$ is equal to a and the modulus of an imaginary number $z = bi$ is equal to b.

Prior Knowledge

You need to be familiar with radians, which are covered in the A level Mathematics book. There is a brief introduction on page 198 of this book.

Figure 6.4 shows the complex number z on an Argand diagram. The length r represents the modulus of the complex number and the angle θ is called the argument of the complex number.

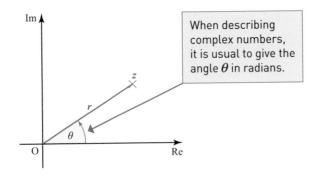

> When describing complex numbers, it is usual to give the angle θ in radians.

Figure 6.4

The argument is measured anticlockwise from the positive real axis. By convention the argument is measured in radians.

However, this angle is not uniquely defined since adding any multiple of 2π to θ gives the same direction. To avoid confusion, it is usual to choose that value of θ for which $-\pi < \theta \leq \pi$, as shown in Figure 6.5.

This is called the **principal argument** of z and is denoted by $\arg(z)$. Every complex number except zero has a unique principal argument.

> The argument of zero is undefined.

Discussion point

→ For the complex number $z = x + yi$, is it true that $\arg(z)$ is given by $\arctan\left(\dfrac{y}{x}\right)$?

Figure 6.5 shows the complex numbers $z_1 = 2 - 3i$ and $z_2 = -2 + 3i$. For both z_1 and z_2, $\dfrac{y}{x} = -\dfrac{3}{2}$ and a calculator gives $\arctan\left(-\dfrac{3}{2}\right) = -0.983$ rad (3 s.f.).

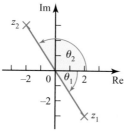

Figure 6.5

The argument of z_1 is the angle θ_1 and this is indeed -0.98 radians. However, the argument of z_2 is the angle θ_2 which is in the second quadrant. It is given by $\pi - 0.98 = 2.16$ radians.

Always draw a diagram when finding the argument of a complex number. This tells you in which quadrant the complex number lies.

Example 6.2

For each of these complex numbers, find the argument of the complex number, giving your answers in radians in exact form or to 3 significant figures as appropriate.

(i) $z_1 = -5 + i$ (ii) $z_2 = 2\sqrt{3} - 2i$ (iii) $z_3 = -5 - 5i$ (iv) $z_4 = -4i$

Solution

(i) $z_1 = -5 + i$

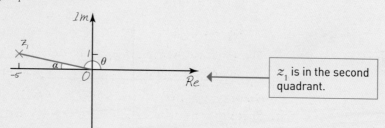

z_1 is in the second quadrant.

Figure 6.6

$$\alpha = \arctan\left(\frac{1}{5}\right) = 0.1973\ldots$$

so $\arg(z_1) = \pi - 0.1973\ldots = 2.94\,(3\text{s.f.})$

(ii) $z_2 = 2\sqrt{3} - 2i$

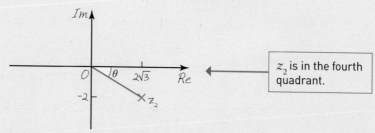

z_2 is in the fourth quadrant.

Figure 6.7

$$\theta = \arctan\left(\frac{2}{2\sqrt{3}}\right) = \frac{\pi}{6}$$

As it is measured in a clockwise direction,

$$\arg(z_2) = -\frac{\pi}{6}.$$

(iii) $z_3 = -5 - 5i$

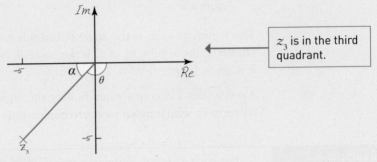

Figure 6.8

$$\alpha = \arctan\left(\frac{5}{5}\right) = \frac{\pi}{4}$$

$$\text{So, } \theta = \pi - \frac{\pi}{4} = \frac{3\pi}{4}$$

Since it is measured in a clockwise direction,
$$\arg(z_3) = -\frac{3\pi}{4}.$$

(iv) $z_4 = -4i$

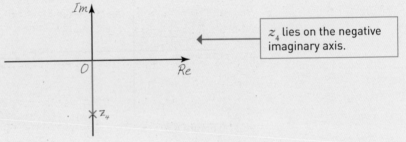

Figure 6.9

On the negative imaginary axis, the argument is $-\frac{\pi}{2}$

$$\arg(z_4) = -\frac{\pi}{2}.$$

The modulus–argument form of a complex number

In Figure 6.10, you can see the relationship between the components of a complex number and its modulus and argument.

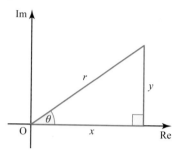

Figure 6.10

Using trigonometry, you can see that $\sin\theta = \dfrac{y}{r}$ and so $y = r\sin\theta$.

Similarly, $\cos\theta = \dfrac{x}{r}$ so $x = r\cos\theta$.

Therefore, the complex number $z = x + y\mathrm{i}$ can be written

$$z = r\cos\theta + r\sin\theta\,\mathrm{i}$$

or

$$z = r\left(\cos\theta + \mathrm{i}\sin\theta\right).$$

> The modulus–argument form of a complex number is sometimes called the *polar* form, as the modulus of a complex number is its distance from the origin, which is also called the *pole*.

This is called the **modulus–argument form** of the complex number and is sometimes written as (r, θ).

You may have noticed in the earlier calculations that values of sin, cos and tan for some angles are exact and can be expressed in surds. You will see these values in the following activity – they are worth memorising as this will help make some calculations quicker.

ACTIVITY 6.1

Copy and complete this table. Use the diagrams in Figure 6.11 to help you.

Give your answers as exact values (involving surds where appropriate), rather than as decimals.

	$\dfrac{\pi}{6}$	$\dfrac{\pi}{4}$	$\dfrac{\pi}{3}$
sin			
cos			
tan			

Table 6.1

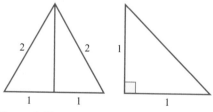

Figure 6.11

ACTIVITY 6.2

Most calculators can convert complex numbers given in the form (x, y) to the form (r, θ) (called *rectangular to polar*, and often shown as $R \rightarrow P$) and from (r, θ) to (x, y) (*polar to rectangular*, $P \rightarrow R$).

Find out how to use these facilities on your calculator.

Does your calculator always give the correct θ, or do you sometimes have to add or subtract 2π?

Example 6.3

Write the following complex numbers in modulus–argument form.

(i) $z_1 = \sqrt{3} + 3i$ (ii) $z_2 = -3 + \sqrt{3}i$

(iii) $z_3 = \sqrt{3} - 3i$ (iv) $z_4 = -3 - \sqrt{3}i$

Solution

Figure 6.12 shows the four complex numbers z_1, z_2, z_3 and z_4.

For each complex number, the modulus is $\sqrt{\left(\sqrt{3}\right)^2 + 3^2} = 2\sqrt{3}$

$$\alpha_1 = \arctan\left(\frac{3}{\sqrt{3}}\right) = \frac{\pi}{3}$$

$$\Rightarrow \arg(z_1) = \frac{\pi}{3}, \text{ so } z_1 = 2\sqrt{3}\left(\cos\frac{\pi}{3} + i\sin\frac{\pi}{3}\right)$$

By symmetry, $\arg(z_3) = -\frac{\pi}{3}$, so $z_3 = 2\sqrt{3}\left(\cos\left(-\frac{\pi}{3}\right) + i\sin\left(-\frac{\pi}{3}\right)\right)$

$$\alpha_2 = \arctan\left(\frac{\sqrt{3}}{3}\right) = \frac{\pi}{6}$$

$$\Rightarrow \arg(z_2) = \pi - \frac{\pi}{6} = \frac{5\pi}{6}, \text{ so } z_2 = 2\sqrt{3}\left(\cos\frac{5\pi}{6} + i\sin\frac{5\pi}{6}\right)$$

By symmetry, $\arg(z_4) = -\frac{5\pi}{6}$, so $z_4 = 2\sqrt{3}\left(\cos\left(-\frac{5\pi}{6}\right) + i\sin\left(-\frac{5\pi}{6}\right)\right)$

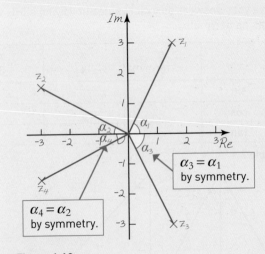

Figure 6.12

① The Argand diagram in Figure 6.13 shows three complex numbers.

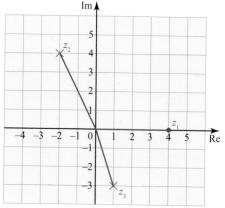

Figure 6.13

Write each of the numbers z_1, z_2 and z_3 in the form:

(i) $a + b$i

(ii) $r(\cos\theta + \mathrm{i}\sin\theta)$, giving answers exactly or to 3 significant figures where appropriate.

② Find the modulus and argument of each of the following complex numbers, giving your answer exactly or to 3 significant figures where appropriate.

(i) $3 + 2\mathrm{i}$ (ii) $-5 + 2\mathrm{i}$ (iii) $-3 - 2\mathrm{i}$ (iv) $2 - 5\mathrm{i}$

③ Find the modulus and argument of each of the complex numbers on this Argand diagram.

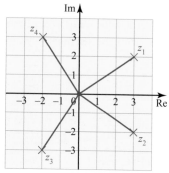

Figure 6.14

Describe the transformations that map z_1 onto each of the other points on the diagram.

④ Write each of the following complex numbers in the form $x + y$i, giving surds in your answer where appropriate.

(i) $4\left(\cos\left(-\dfrac{\pi}{2}\right) + \mathrm{i}\sin\left(-\dfrac{\pi}{2}\right)\right)$

(ii) $7\left(\cos\dfrac{3\pi}{4} + \mathrm{i}\sin\dfrac{3\pi}{4}\right)$

(iii) $\quad 3\left(\cos\dfrac{5\pi}{6} + \mathrm{i}\sin\dfrac{5\pi}{6}\right)$

(iv) $\quad 5\left(\cos\left(-\dfrac{\pi}{6}\right) + \mathrm{i}\sin\left(-\dfrac{\pi}{6}\right)\right)$

⑤ For each complex number, find the modulus and argument, and hence write the complex number in modulus-argument form.

Give the argument in radians, either as a multiple of π or correct to 3 significant figures.

(i) $\quad 1$ (ii) $\quad -2$ (iii) $\quad 3\mathrm{i}$ (iv) $\quad -4\mathrm{i}$

⑥ For each of the complex numbers below, find the modulus and argument, and hence write the complex number in modulus-argument form.

Give the argument in radians as a multiple of π.

(i) $\quad 1 + \mathrm{i}$ (ii) $\quad -1 + \mathrm{i}$ (iii) $\quad -1 - \mathrm{i}$ (iv) $\quad 1 - \mathrm{i}$

⑦ For each complex number, find the modulus and principal argument, and hence write the complex number in modulus-argument form.

Give the argument in radians, either as a multiple of π or correct to 3 significant figures.

(i) $\quad 6\sqrt{3} + 6\mathrm{i}$ (ii) $\quad 3 - 4\mathrm{i}$ (iii) $\quad -12 + 5\mathrm{i}$

(iv) $\quad 4 + 7\mathrm{i}$ (v) $\quad -58 - 93\mathrm{i}$

⑧ Express each of these complex numbers in the form $r(\cos\theta + \mathrm{i}\sin\theta)$ giving the argument in radians, either as a multiple of π or correct to 3 significant figures.

(i) $\quad \dfrac{2}{3 - \mathrm{i}}$ (ii) $\quad \dfrac{3 - 2\mathrm{i}}{3 - \mathrm{i}}$ (iii) $\quad \dfrac{-2 - 5\mathrm{i}}{3 - \mathrm{i}}$

⑨ Represent each of the following complex numbers on a separate Argand diagram and write it in the form $x + y\mathrm{i}$, giving surds in your answer where appropriate.

(i) $\quad |z| = 2,\ \arg(z) = \dfrac{\pi}{2}$ (ii) $\quad |z| = 3,\ \arg(z) = \dfrac{\pi}{3}$

(iii) $\quad |z| = 7,\ \arg(z) = \dfrac{5\pi}{6}$ (iv) $\quad |z| = 1,\ \arg(z) = -\dfrac{\pi}{4}$

(v) $\quad |z| = 5,\ \arg(z) = -\dfrac{2\pi}{3}$ (vi) $\quad |z| = 6,\ \arg(z) = -2$

⑩ Given that $\arg(5 + 2\mathrm{i}) = \alpha$, find the argument of each of the following in terms of α.

(i) $\quad -5 - 2\mathrm{i}$ (ii) $\quad 5 - 2\mathrm{i}$ (iii) $\quad -5 + 2\mathrm{i}$

(iv) $\quad 2 + 5\mathrm{i}$ (v) $\quad -2 + 5\mathrm{i}$

⑪ The complex numbers z_1 and z_2 are shown on the Argand diagram in Figure 6.15.

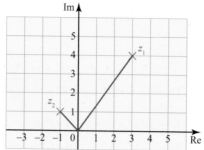

Figure 6.15

(i) Find the modulus and argument of each of the two numbers.

(ii) (a) Find $z_1 z_2$ and $\dfrac{z_1}{z_2}$.

(b) Find the modulus and argument of each of $z_1 z_2$ and $\dfrac{z_1}{z_2}$.

(iii) What rules can you deduce about the modulus and argument of the two complex numbers and the answers to part (ii)(b)?

2 Multiplying and dividing complex numbers in modulus-argument form

Prior knowledge

You need to be familiar with the compound angle formulae. These are covered in the A level Mathematics book, and a brief introduction is given on page 201 of this book.

ACTIVITY 6.3

What is the geometrical effect of multiplying one complex number by another? To explore this question, start with the numbers $z_1 = 2 + 3\mathrm{i}$ and $z_2 = \mathrm{i}z_1$.

(i) Plot the vectors z_1 and z_2 on the same Argand diagram, and describe the geometrical transformation that maps the vector z_1 to the vector z_2.

(ii) Repeat part (i) with $z_1 = 2 + 3\mathrm{i}$ and $z_2 = 2\mathrm{i}z_1$.

(iii) Repeat part (i) with $z_1 = 2 + 3\mathrm{i}$ and $z_2 = (1 + \mathrm{i})z_1$.

You will have seen in Activity 6.3 that multiplying one complex number by another involves a combination of an enlargement and a rotation.

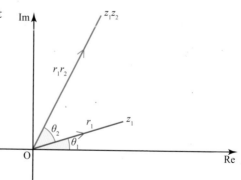

Figure 6.16

You obtain the vector z_1z_2 by enlarging the vector z_1 by the scale factor $|z_2|$, and rotate it anticlockwise through an angle of $\arg(z_2)$.

So to multiply complex numbers in modulus-argument form, you *multiply* their moduli and *add* their arguments.

$$|z_1z_2| = |z_1||z_2|$$

$$\arg(z_1z_2) = \arg(z_1) + \arg(z_2)$$

> You may need to add or subtract 2π to give the principal argument.

You can prove these results using the compound angle formulae.

$$z_1z_2 = r_1(\cos\theta_1 + i\sin\theta_1) \times r_2(\cos\theta_2 + i\sin\theta_2)$$
$$= r_1r_2(\cos\theta_1\cos\theta_2 + i\cos\theta_1\sin\theta_2 + i\sin\theta_1\cos\theta_2 - \sin\theta_1\sin\theta_2)$$
$$= r_1r_2[(\cos\theta_1\cos\theta_2 - \sin\theta_1\sin\theta_2) + i(\cos\theta_1\sin\theta_2 + \sin\theta_1\cos\theta_2)]$$
$$= r_1r_2[(\cos(\theta_1 + \theta_2)) + i\sin(\theta_1 + \theta_2)]$$

> The identity $\cos(\theta_1 + \theta_2)$.

> The identity $\sin(\theta_1 + \theta_2)$.

So, $|z_1z_2| = r_1r_2$ and $\arg(z_1z_2) = \theta_1 + \theta_2$.

Dividing complex numbers works in a similar way. You obtain the vector $\dfrac{z_1}{z_2}$ by enlarging the vector z_1 by the scale factor $\dfrac{1}{|z_2|}$, and rotate it *clockwise* through an angle of $\arg(z_2)$.

> This is equivalent to rotating it anticlockwise through an angle of $-\arg(z_2)$.

So, to divide complex numbers in modulus-argument form, you *divide* their moduli and *subtract* their arguments.

$$\left|\frac{z_1}{z_2}\right| = \frac{|z_1|}{|z_2|}$$

$$\arg\left(\frac{z_1}{z_2}\right) = \arg(z_1) - \arg(z_2)$$

You can prove this easily from the multiplication results by letting $\dfrac{z_1}{z_2} = w$, so that $z_1 = wz_2$.

Then $|z_1| = |w||z_2|$, so $|w| = \dfrac{|z_1|}{|z_2|}$

and $\arg(z_1) = \arg(w) + \arg(z_2)$, so $\arg(w) = \arg(z_1) - \arg(z_2)$.

Example 6.4

The complex numbers w and z are given by $w = 2\left(\cos\dfrac{\pi}{4} + i\sin\dfrac{\pi}{4}\right)$ and $z = 5\left(\cos\dfrac{5\pi}{6} + i\sin\dfrac{5\pi}{6}\right)$.

Find (i) wz and (ii) $\dfrac{w}{z}$ in modulus-argument form. Illustrate each of these on a separate Argand diagram.

Solution

$|w| = 2 \quad \arg(w) = \dfrac{\pi}{4}$

$|z| = 5 \quad \arg(w) = \dfrac{5\pi}{6}$

(i) $|wz| = |w||z| = 2 \times 5 = 10$

$\arg(w) + \arg(z) = \dfrac{\pi}{4} + \dfrac{5\pi}{6} = \dfrac{13\pi}{12}$

> This is not in the range $-\pi < \theta \leqslant \pi$.

so $\arg(wz) = \dfrac{13\pi}{12} - 2\pi = -\dfrac{11}{12}\pi$

> Subtract 2π to obtain the principal argument.

$$wz = 10\left(\cos\left(-\dfrac{11\pi}{12}\right) + i\sin\left(-\dfrac{11\pi}{12}\right)\right)$$

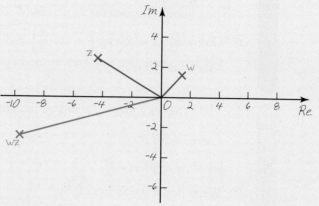

Figure 6.17

(ii) $\left|\dfrac{w}{z}\right| = \dfrac{|w|}{|z|} = \dfrac{2}{5}$

$\arg(w) - \arg(z) = \dfrac{\pi}{4} - \dfrac{5\pi}{6} = -\dfrac{7\pi}{12}$

$$\dfrac{w}{z} = \dfrac{2}{5}\left(\cos\left(-\dfrac{7\pi}{12}\right) + i\sin\left(-\dfrac{7\pi}{12}\right)\right)$$

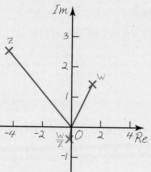

Figure 6.18

① The complex numbers w and z shown in the Argand diagram are $w = 1 + i$ and $z = 1 - \sqrt{3}i$

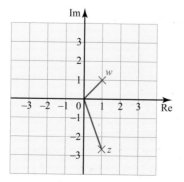

Figure 6.19

    (i)   Find the modulus and argument of each of the complex numbers w and z.

    (ii)   Hence write down the modulus and argument of

       (a) wz

       (b) $\dfrac{w}{z}$

    (iii) Show the points w, z, wz and $\dfrac{w}{z}$ on a copy of the Argand diagram.

② Given that $z = 2\left(\cos\dfrac{\pi}{4} + \mathrm{i}\sin\dfrac{\pi}{4}\right)$ and $w = 3\left(\cos\dfrac{\pi}{3} + \mathrm{i}\sin\dfrac{\pi}{3}\right)$, find the following complex numbers in modulus-argument form

    (i) wz (ii) $\dfrac{w}{z}$ (iii) $\dfrac{z}{w}$ (iv) $\dfrac{1}{z}$

③ The complex numbers z and w are defined as follows:

$z = -3 + 3\sqrt{3}\mathrm{i}$

$|w| = 18,\ \arg(w) - \dfrac{\pi}{6}$

    Write down the values of

    (i) $\arg(z)$ (ii) $|z|$ (iii) $\arg(zw)$ (iv) $|zw|$.

④ Given that $z = 6\left(\cos\dfrac{\pi}{6} + \mathrm{i}\sin\dfrac{\pi}{6}\right)$ and $w = 2\left(\cos\left(-\dfrac{\pi}{4}\right) + \mathrm{i}\sin\left(-\dfrac{\pi}{4}\right)\right)$, find the following complex numbers in modulus-argument form:

    (i) w^2 (ii) z^5 (iii) $w^3 z^4$

    (iv) $5\mathrm{i}z$ (v) $(1 + \mathrm{i})w$

⑤ Find the multiplication scale factor and the angle of rotation which maps

    (i) the vector $2 + 3\mathrm{i}$ to the vector $5 - 2\mathrm{i}$

    (ii) the vector $-4 + \mathrm{i}$ to the vector $3\mathrm{i}$.

⑥ Prove that, in general, $\arg\left[\dfrac{1}{z}\right] = -\arg[z]$. What are the exceptions to this rule?

⑦ (i) Find the real and imaginary parts of $\dfrac{-1 + \mathrm{i}}{1 + \sqrt{3}\mathrm{i}}$.

    (ii) Express $-1 + \mathrm{i}$ and $1 + \sqrt{3}\mathrm{i}$ in modulus-argument form.

    (iii) Hence show that $\cos\dfrac{5\pi}{12} = \dfrac{\sqrt{3} - 1}{2\sqrt{2}}$, and find an exact expression for $\sin\dfrac{5\pi}{12}$.

⑧ Prove that for three complex numbers $w = r_1(\cos\theta_1 + \mathrm{i}\sin\theta_1)$, $z = r_2(\cos\theta_2 + \mathrm{i}\sin\theta_2)$ and $p = r_3(\cos\theta_3 + \mathrm{i}\sin\theta_3)$, $|wzp| = |w||z||p|$ and $\arg(wzp) = \arg(w) + \arg(z) + \arg(p)$.

3 Loci in the Argand diagram

A locus is the set of locations that a point can occupy when constrained by a given rule. The plural of locus is loci.

Loci of the form $|z - a| = r$

Figure 6.20 shows the positions for two general complex numbers $z_1 = x_1 + y_1 i$ and $z_2 = x_2 + y_2 i$.

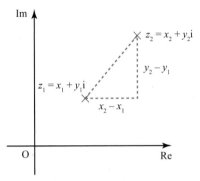

Figure 6.20

You saw earlier that the complex number $z_2 - z_1$ can be represented by the vector from the point representing z_1 to the point representing z_2 (see Figure 6.20). This is the key to solving many questions about sets of points in an Argand diagram, as shown in the following example.

Example 6.5

Draw Argand diagrams showing the following sets of points z for which

(i) $|z| = 5$

(ii) $|z - 3| = 5$

(iii) $|z - 4i| = 5$

(iv) $|z - 3 - 4i| = 5$

Solution

(i) $|z| = 5$

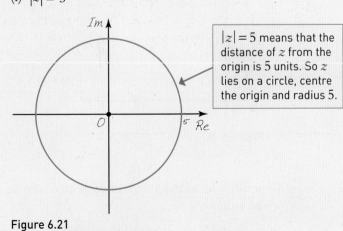

$|z| = 5$ means that the distance of z from the origin is 5 units. So z lies on a circle, centre the origin and radius 5.

Figure 6.21

(ii) $|z - 3| = 5$

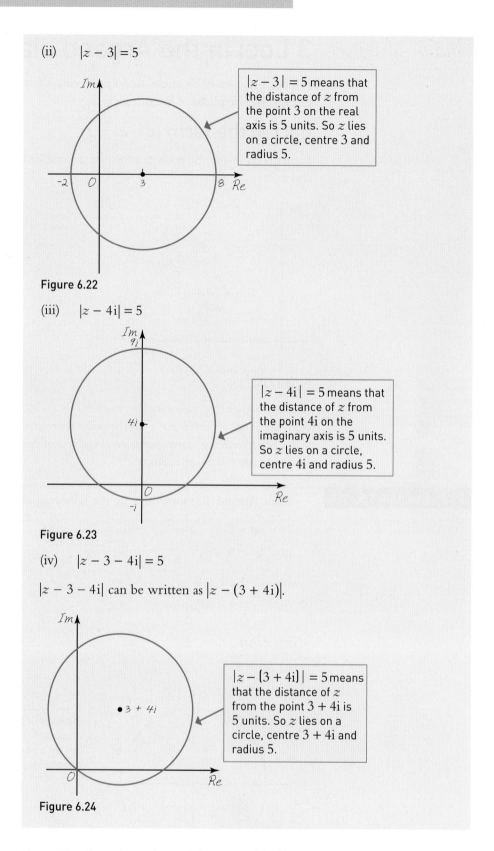

> $|z - 3| = 5$ means that the distance of z from the point 3 on the real axis is 5 units. So z lies on a circle, centre 3 and radius 5.

Figure 6.22

(iii) $|z - 4i| = 5$

> $|z - 4i| = 5$ means that the distance of z from the point 4i on the imaginary axis is 5 units. So z lies on a circle, centre 4i and radius 5.

Figure 6.23

(iv) $|z - 3 - 4i| = 5$

$|z - 3 - 4i|$ can be written as $|z - (3 + 4i)|$.

> $|z - (3 + 4i)| = 5$ means that the distance of z from the point $3 + 4i$ is 5 units. So z lies on a circle, centre $3 + 4i$ and radius 5.

Figure 6.24

Generally, a locus in an Argand diagram of the form $|z - a| = r$ is a circle, centre a and radius r.

In the example above, each locus is the set of points on the circumference of the circle. It is possible to define a region in the Argand diagram in a similar way.

Example 6.6

Draw Argand diagrams showing the following sets of points z for which

(i) $|z| < 5$

(ii) $|z - 3| > 5$

(iii) $|z - 4i| \leqslant 5$

Solution

(i) $|z| < 5$

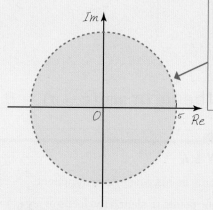

$|z| < 5$ means that all the points inside the circle are included, but not the points on the circumference of the circle. The circle is shown as a dotted line to indicate that it is not part of the locus.

Figure 6.25

(ii) $|z - 3| > 5$

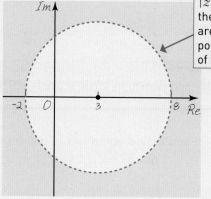

$|z - 3| > 5$ means that all the points outside the circle are included, but not the points on the circumference of the circle.

Figure 6.26

(iii) $|z - 4i| \leqslant 5$

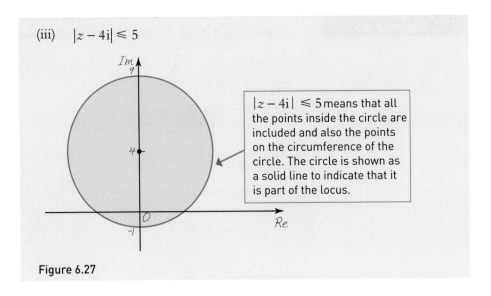

$|z - 4i| \leqslant 5$ means that all the points inside the circle are included and also the points on the circumference of the circle. The circle is shown as a solid line to indicate that it is part of the locus.

Figure 6.27

Loci of the form $\mathbf{arg}(z - a) = \theta$

> ## ACTIVITY 6.4
>
> (i) Plot some points which have argument $\dfrac{\pi}{4}$.
>
> Use your points to sketch the locus of $\arg(z) = \dfrac{\pi}{4}$.
>
> Is the point $-2 - 2i$ on this locus?
>
> How could you describe the locus?
>
> (ii) Which of the following complex numbers satisfy $\arg(z - 2) = \dfrac{\pi}{4}$?
>
> (a) $z = 4$
>
> (b) $z = 3 + i$
>
> (c) $z = 4i$
>
> (d) $z = 8 + 6i$
>
> (e) $z = 1 - i$
>
> Describe and sketch the locus of points which satisfy $\arg(z - 2) = \dfrac{\pi}{4}$.

In Activity 6.4 you looked at the loci of points of the form $\arg(z - a) = \theta$ in the case when $a = 2$ and $\theta = \dfrac{\pi}{4}$. On the Argand diagram the locus looks like this.

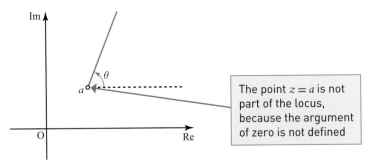

The point $z = a$ is not part of the locus, because the argument of zero is not defined

Figure 6.28

Note that the half line does not include the end point where $z = a$.

The locus is a half line, starting from $z = a$ and making an angle θ measured anticlockwise from the direction of the positive real axis.

Example 6.7

Sketch the locus of z in an Argand diagram when

(i) $\arg(z - 3) = \dfrac{2\pi}{3}$

(ii) $\arg(z + 2i) = \dfrac{\pi}{6}$

(iii) $\arg(z - 1 + 4i) = -\dfrac{\pi}{4}$.

Solution

(i) This is a half line starting from $z = 3$, at an angle $\dfrac{2\pi}{3}$.

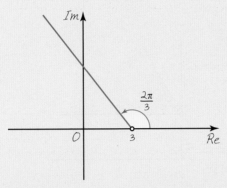

Figure 6.29

(ii) This can be written in the form $\arg\big(z - (-2i)\big) = \dfrac{\pi}{6}$ so it is a half line starting from $-2i$ at an angle $\dfrac{\pi}{6}$.

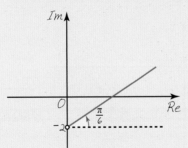

Figure 6.30

(iii) This can be written $\arg\big(z - (1 - 4i)\big) = -\dfrac{\pi}{4}$ so it is a half line starting from $1 - 4i$ at an angle $-\dfrac{\pi}{4}$.

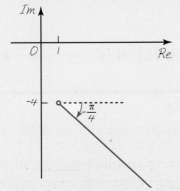

Figure 6.31

Example 6.8

Sketch diagrams that represent the regions represented by

(i) $0 \leqslant \arg(z - 3i) \leqslant \dfrac{\pi}{3}$

(ii) $-\dfrac{\pi}{4} < \arg(z - 3 + 4i) < \dfrac{\pi}{4}$.

Solution

(i) This is the region between the two half lines starting at $z = 3i$, at angle 0 and angle $\dfrac{\pi}{3}$.

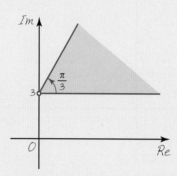

Figure 6.32

(ii) $\arg(z - 3 + 4i)$ can be written $\arg\big(z - (3 - 4i)\big)$ so this is the region between two half lines starting at $3 - 4i$ at angles $-\dfrac{\pi}{4}$ and $\dfrac{\pi}{4}$.

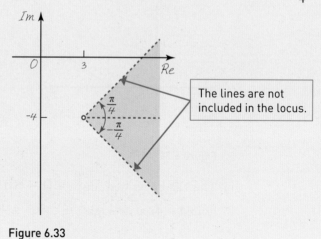

The lines are not included in the locus.

Figure 6.33

Loci of the form $|z - a| = |z - b|$

ACTIVITY 6.5

On an Argand diagram, mark the points $3 + 4i$ and $-1 + 2i$. Identify some points that are the same distance from both points.

Use your diagram to describe and sketch the locus $|z - 3 - 4i| = |z + 1 - 2i|$.

Generally, the locus $|z - a| = |z - b|$ represents the locus of all points which lie on the perpendicular bisector between the points represented by the complex numbers a and b.

Example 6.9

Show each of the following sets of points on an Argand diagram.

(i) $|z - 3 - 4i| = |z + 1 - 2i|$

(ii) $|z - 3 - 4i| < |z + 1 - 2i|$

(iii) $|z - 3 - 4i| \geqslant |z + 1 - 2i|$

Solution

(i) The condition can be written as $|z - (3 + 4i)| = |z - (-1 + 2i)|$.

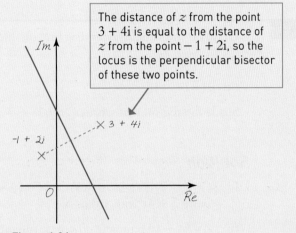

> The distance of z from the point $3 + 4i$ is equal to the distance of z from the point $-1 + 2i$, so the locus is the perpendicular bisector of these two points.

Figure 6.34

(ii) $|z - 3 - 4i| < |z + 1 - 2i|$

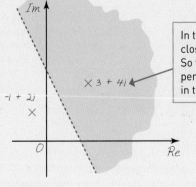

> In this case the locus includes all the points closer to the point $3 + 4i$ than to $-1 + 2i$. So the locus is the shaded area. The perpendicular bisector itself is not included in the locus, so it is shown as a dotted line.

Figure 6.35

(iii) $|z - 3 - 4i| \geqslant |z + 1 - 2i|$

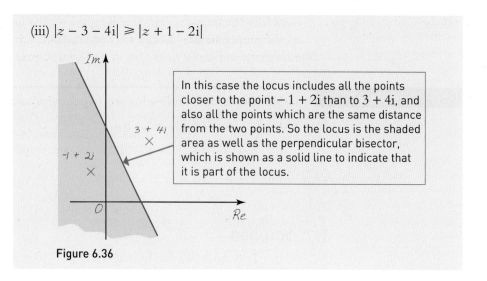

In this case the locus includes all the points closer to the point $-1 + 2i$ than to $3 + 4i$, and also all the points which are the same distance from the two points. So the locus is the shaded area as well as the perpendicular bisector, which is shown as a solid line to indicate that it is part of the locus.

Figure 6.36

Example 6.10

Draw, on the same Argand diagram, the loci

(i) $|z - 3 - 4i| = 5$

(ii) $|z| = |z - 4i|$.

Shade the region that satisfies both $|z - 3 - 4i| \leqslant 5$ and $|z| \leqslant |z - 4i|$.

Solution

(i) $|z - 3 - 4i|$ can be written as $|z - (3 + 4i)|$ so (i) is a circle centre $3 + 4i$ with radius 5.

(ii) $|z| = |z - 4i|$ represents the perpendicular bisector of the line between the points $z = 0$ and $z = 4i$.

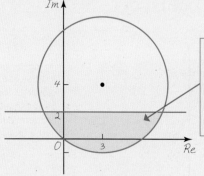

$|z - 3 - 4i| \leqslant 5$ represents the circumference and the inside of the circle. $|z| \leqslant |z - 4i|$ represents the side of the perpendicular bisector that is nearer to the origin including the perpendicular bisector itself. The shaded area represents the region for which both conditions are true.

Figure 6.37

⚠ Don't get confused between loci of the forms $|z - a| = r$ and $|z - a| = |z - b|$.

$|z - a| = r$ represents a circle, centred on the complex number a, with radius r.

$|z - a| = |z - b|$ represents the perpendicular bisector of the line between the points a and b.

Exercise 6.3

① For each of parts (i) to (iv), draw an Argand diagram showing the set of points z for which the given condition is true.

(i) $|z| = 2$

(ii) $|z - 2i| = 2$

(iii) $|z - 2| = 2$

(iv) $\left|z + \sqrt{2} + \sqrt{2}i\right| = 2$

② For each of parts (i) to (iv), draw an Argand diagram showing the set of points z for which the given condition is true.

(i) $\arg(z) = \dfrac{\pi}{3}$

(ii) $\arg\left(z + 1 + \sqrt{3}i\right) = \dfrac{\pi}{3}$

(iii) $\arg\left(z - 1 + \sqrt{3}i\right) = \dfrac{2\pi}{3}$

(iv) $\arg\left(z - 1 - \sqrt{3}i\right) = -\dfrac{2\pi}{3}$

③ For each of parts (i) to (iv), draw an Argand diagram showing the set of points z for which the given condition is true.

(i) $|z - 8| = |z - 4|$

(ii) $|z - 2 - 4i| = |z - 6 - 8i|$

(iii) $|z + 5 - 2i| = |z + 3i|$

(iv) $|z + 3 + 5i| = |z - i|$

④ Write down the loci for the sets of points z that are represented in each of these Argand diagrams.

(i)

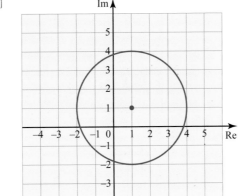

Figure 6.38

(ii)

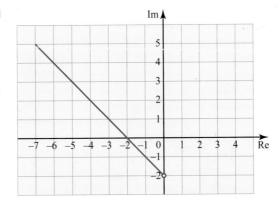

Figure 6.39

(iii)

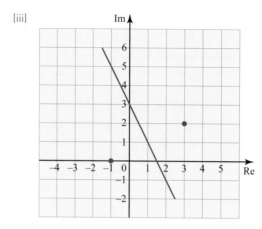

Figure 6.40

⑤ Write down, in terms of z, the loci for the regions that are represented in each of these Argand diagrams.

(i)

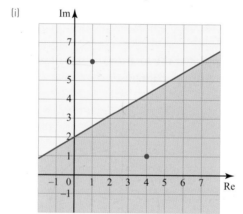

Figure 6.41

(ii)

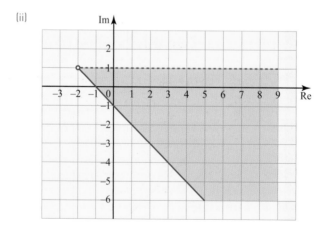

Figure 6.42

(iii)

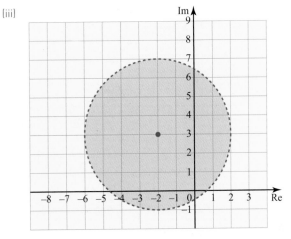

Figure 6.43

⑥ Draw an Argand diagram showing the set of points z for which $|z - 12 + 5i| \leqslant 7$. Use the diagram to prove that, for these z, $6 \leqslant |z| \leqslant 20$.

⑦ For each of parts (i) to (iii), draw an Argand diagram showing the set of points z for which the given condition is true.

(i) $\arg(z - 3 + i) \leqslant -\dfrac{\pi}{6}$

(ii) $0 \leqslant \arg(z - 3i) \leqslant \dfrac{3\pi}{4}$

(iii) $-\dfrac{\pi}{4} < \arg(z + 5 - 3i) < \dfrac{\pi}{3}$

⑧ On an Argand diagram shade in the regions represented by the following inequalities.

(i) $|z - 3| \leqslant 2$

(ii) $|z - 6i| > |z + 2i|$

(iii) $2 \leqslant |z - 3 - 4i| \leqslant 4$

(iv) $|z + 3 + 6i| \leqslant |z - 2 - 7i|$.

⑨ Shade on an Argand diagram the region satisfied by the inequalities $|z - 1 + i| \leqslant 1$ and $-\dfrac{\pi}{3} < \arg(z) < 0$.

⑩

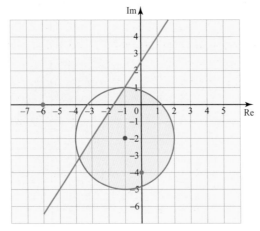

Figure 6.44

(i) For this Argand diagram, write down in terms of z

    (a) the loci of the set of points on the circle

    (b) the loci of the set of points on the straight line.

(ii) Using inequalities, express in terms of z the shaded region on the Argand diagram.

⑪ Sketch on the same Argand diagram

    (i) the locus of points $|z - 2 + 2i| = 3$

    (ii) the locus of points $\arg(z - 2 + 2i) = -\dfrac{\pi}{4}$

    (iii) the locus of points $\arg(z - 2 + 2i) = \dfrac{\pi}{2}$.

Shade the region defined by the inequalities $|z - 2 + 2i| \leqslant 3$ $\arg(z - 2 + 2i) \geqslant -\dfrac{\pi}{4}$ and $\arg(z - 2 + 2i) \leqslant \dfrac{\pi}{2}$.

⑫ You are given the complex number $w = -\sqrt{3} + 3i$.

    (i) Find $\arg(w)$ and $|w - 2i|$.

    (ii) On an Argand diagram, shade the region representing complex numbers z which satisfy both of these inequalities:

$$|z - 2i| \leqslant 2 \text{ and } \frac{\pi}{2} \leqslant \arg z \leqslant \frac{2\pi}{3}$$

Indicate the point on your diagram which corresponds to w.

⑬ Sketch a diagram that represents the region represented by

$$|z - 2 - 2i| \leqslant 2 \text{ \textbf{and} } 0 \leqslant \arg(z - 2i) \leqslant \frac{\pi}{4}.$$

⑭ By using an Argand diagram, determine if it is possible to find values of z for which $|z - 2 + i| \geqslant 10$ and $|z + 4 + 2i| \leqslant 2$ simultaneously.

⑮ What are the greatest and least values of $|z + 3 - 2i|$ if $|z - 5 + 4i| \leqslant 3$?

⑯ You are given that $|z - 3| = 2|z - 3 + 9i|$.

    (i) Show, using algebra with $z = x + yi$, that the locus of z is a circle and state the centre and radius of the circle.

    (ii) Sketch the locus of the circle on an Argand diagram.

LEARNING OUTCOMES

When you have completed this chapter you should be able to:

➤ find the modulus of a complex number

➤ find the principal argument of a complex number using radians

➤ express a complex number in modulus-argument form

➤ multiply and divide complex numbers in modulus-argument form

➤ represent multiplication and division of two complex numbers on an Argand diagram

➤ represent and interpret sets of complex numbers as loci on an Argand diagram:

  ➤ circles of the form $|z - a| = r$

  ➤ half-lines of the form $\arg(z - a) = \theta$

  ➤ lines of the form $|z - a| = |z - b|$

➤ represent and interpret regions defined by inequalities based on the above.

KEY POINTS

1 The modulus of $z = x + y\mathrm{i}$ is $|z| = \sqrt{x^2 + y^2}$. This is the distance of the point z from the origin on the Argand diagram.

2 The argument of z is the angle θ, measured in radians, between the line connecting the origin and the point z and the positive real axis.

3 The principal argument of z, arg (z), is the angle θ, measured in radians, for which $-\pi < \theta \leqslant \pi$, between the line connecting the origin and the point z and the positive real axis.

4 For a complex number z, $zz* = |z|^2$.

5 The modulus–argument form of z is $z = r\left(\cos\theta + \mathrm{i}\sin\theta\right)$, where $r = |z|$ and $\theta = \arg(z)$. This is often written as (r, θ)

6 For two complex numbers z_1 and z_2:

$$|z_1 z_2| = |z_1||z_2| \qquad \arg\left(z_1 z_2\right) = \arg\left(z_1\right) + \arg\left(z_2\right)$$

$$\left|\frac{z_1}{z_2}\right| = \frac{|z_1|}{|z_2|} \qquad \arg\left(\frac{z_1}{z_2}\right) = \arg\left(z_1\right) - \arg\left(z_2\right)$$

7 The distance between the points z_1 and z_2 in an Argand diagram is $|z_1 - z_2|$.

8 $|z - a| = r$ represents a circle, centre a and radius r.

$|z - a| < r$ represents the interior of the circle, and $|z - a| > r$ represents the exterior of the circle.

9 $\arg(z - a) = \theta$ represents a half line starting at $z = a$ at an angle of θ from the positive real direction.

10 $|z - a| = |z - b|$ represents the perpendicular bisector of the points a and b.

FUTURE USES

■ Work on complex numbers will be developed further in A level Further Mathematics.

■ Complex numbers will be needed for work on differential equations in A level Further Mathematics, in particular in modelling oscillations (simple harmonic motion).

Matrices and their inverses

Figure 7.1 Sierpinsky triangle.

Discussion point

→ What is the same about each of the triangles in the diagram? How many of the yellow triangles are needed to cover the large purple triangle?

The diagram in Figure 7.1 is called a Sierpinsky triangle. The pattern can be continued with smaller and smaller triangles.

1 The determinant of a matrix

Figure 7.2 shows the unit square, labelled OIPJ, and the parallelogram OI'P'J' formed when OIPJ is transformed using the matrix $\begin{pmatrix} 5 & 4 \\ 1 & 2 \end{pmatrix}$.

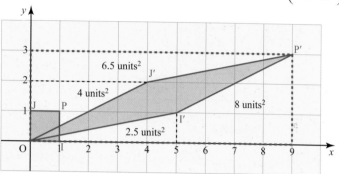

Figure 7.2

What effect does the transformation have on the area of OIPJ?

The area of OIPJ is 1 unit².

To find the area of OI'P'J', a rectangle has been drawn surrounding it. The area of the rectangle is $9 \times 3 = 27$ units². The part of the rectangle that is not inside OI'P'J' has been divided up into two triangles and two trapezia and their areas are shown on the diagram.

So, area OI'P'J' $= 27 - 2.5 - 8 - 6.5 - 4 = 6$ units².

The interesting question is whether you could predict this answer from the numbers in the matrix $\begin{pmatrix} 5 & 4 \\ 1 & 2 \end{pmatrix}$.

You can see that $5 \times 2 - 4 \times 1 = 6$. Is this just a coincidence?

To answer that question you need to transform the unit square by the general 2×2 matrix $\begin{pmatrix} a & b \\ c & d \end{pmatrix}$ and see whether the area of the transformed figure is $(ad - bc)$ units². The answer is, 'Yes', and the proof is left for you to do in the activity below.

ACTIVITY 7.1

The unit square is transformed by the matrix $\begin{pmatrix} a & b \\ c & d \end{pmatrix}$.

Prove that the resulting shape is a parallelogram with area $(ad - bc)$ units².

It is now evident that the quantity $(ad - bc)$ is the area scale factor associated with the transformation matrix $\begin{pmatrix} a & b \\ c & d \end{pmatrix}$. It is called the **determinant** of the matrix.

The are several ways to denote the determinant of a matrix **M**: det (**M**), det **M**, $|\mathbf{M}|$ and Δ.

Example 7.1

A shape S has area $8\,\mathrm{cm}^2$. S is mapped to a shape T under the transformation represented by the matrix $\mathbf{M} = \begin{pmatrix} 1 & -2 \\ 3 & 0 \end{pmatrix}$.

Find the area of shape T.

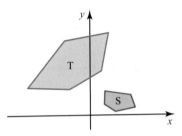

Figure 7.3

> **Note**
>
> In Example 7.1, it does not matter what shape S looks like; for any shape S with area $8\,\mathrm{cm}^2$, the area of the image T will always be $48\,\mathrm{cm}^2$.

Solution

$$\det\begin{pmatrix} 1 & -2 \\ 3 & 0 \end{pmatrix} = (1 \times 0) - (-2 \times 3) = 0 + 6 = 6$$

The area scale factor of the transformation is 6 ...

Area of T $= 8 \times 6$

... and so the area of the original shape is multiplied by 6.

$$= 48\,\mathrm{cm}^2$$

Example 7.2

(i) Draw a diagram to show the image of the unit square OIPJ under the transformation represented by the matrix $\mathbf{M} = \begin{pmatrix} 2 & 3 \\ 4 & 1 \end{pmatrix}$.

(ii) Find det $\mathbf{M}$.

(iii) Use your answer to part (ii) to find the area of the transformed shape.

Solution

(i) $\begin{pmatrix} 2 & 3 \\ 4 & 1 \end{pmatrix}\begin{pmatrix} 0 & 1 & 1 & 0 \\ 0 & 0 & 1 & 1 \end{pmatrix} = \begin{pmatrix} 0 & 2 & 5 & 3 \\ 0 & 4 & 5 & 1 \end{pmatrix}$

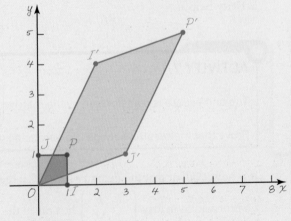

Figure 7.4

(ii) $\det \begin{pmatrix} 2 & 3 \\ 4 & 1 \end{pmatrix} = (2 \times 1) - (3 \times 4) = 2 - 12 = -10$

(iii) The area of the transformed shape is 10 square units.

> Notice that the determinant is negative. Since area cannot be negative, the area of the transformed shape is 10 square units.

The sign of the determinant does have significance. If you move anticlockwise around the original unit square you come to vertices O, I, P, J in that order. However, moving anticlockwise about the image reverses the order of the vertices i.e. O, J′, P′, I′. This reversal in the order of the vertices produces the negative determinant.

Discussion point

→ Which of the following transformations reverse the order of the vertices:
(i)   rotation
(ii) reflection
(iii) enlargement?

Check your answers by finding the determinants of matrices representing these transformations.

Example 7.3

Given that $\mathbf{P} = \begin{pmatrix} 2 & 1 \\ 0 & 1 \end{pmatrix}$ and $\mathbf{Q} = \begin{pmatrix} 2 & 1 \\ 1 & 2 \end{pmatrix}$, find

(i)   det $\mathbf{P}$

(ii)   det $\mathbf{Q}$

(iii) det ($\mathbf{PQ}$).

What do you notice?

Solution

(i)   det $\mathbf{P} = 2 - 0 = 2$

(ii)   det $\mathbf{Q} = 4 - 1 = 3$

(iii) $\mathbf{PQ} = \begin{pmatrix} 2 & 1 \\ 0 & 1 \end{pmatrix}\begin{pmatrix} 2 & 1 \\ 1 & 2 \end{pmatrix} = \begin{pmatrix} 5 & 4 \\ 1 & 2 \end{pmatrix}$ det ($\mathbf{PQ}$) = 10 − 4 = 6

> Remember that a transformation **MN** means 'apply **N**, then apply **M**'.

The determinant of $\mathbf{PQ}$ is given by det $\mathbf{P} \times$ det $\mathbf{Q}$.

The example above illustrates the general result that det ($\mathbf{MN}$) = det $\mathbf{M} \times$ det $\mathbf{N}$.

> 🖥 **TECHNOLOGY**
>
> You can use your calculator to find the determinant of 3×3 matrices.

This result makes sense in terms of transformations. In Example 7.3, applying $\mathbf{Q}$ involves an area scale factor of 3, and applying $\mathbf{P}$ involves an area scale factor of 2. So applying $\mathbf{Q}$ followed by $\mathbf{P}$, represented by the matrix $\mathbf{PQ}$, involves an area scale factor of 6.

The work so far has been restricted to 2×2 matrices. All square matrices have determinants; for a 3×3 matrix the determinant represents a volume scale factor. However, a non-square matrix does not have a determinant.

Discussion points

→ Find out how to calculate the determinant of square matrices using your calculator.

→ Use your calculator to find the determinant of the matrix $\begin{pmatrix} 2 & 0 & 0 \\ 0 & 2 & 0 \\ 0 & 0 & 2 \end{pmatrix}$.

→ Describe the transformation represented by the matrix $\begin{pmatrix} 2 & 0 & 0 \\ 0 & 2 & 0 \\ 0 & 0 & 2 \end{pmatrix}$ and explain the significance of the determinant.

You will learn how to find the determinant of a 3×3 matrix without a calculator in the next section.

ACTIVITY 7.2

Using the matrices $\mathbf{A} = \begin{pmatrix} 1 & 4 \\ -2 & 3 \end{pmatrix}$ and $\mathbf{B} = \begin{pmatrix} 3 & 1 \\ 2 & 2 \end{pmatrix}$, verify that

$(\mathbf{AB})^{-1} = \mathbf{B}^{-1}\mathbf{A}^{-1}$

Example 7.4

A transformation is represented by the matrix $\mathbf{A} = \begin{pmatrix} -1 & 0 & 0 \\ 0 & 1 & 0 \\ 0 & 0 & 1 \end{pmatrix}$.

(i) Describe the transformation represented by $\mathbf{A}$.

(ii) Using a calculator, find the determinant of $\mathbf{A}$.

(iii) Decide whether the transformation represented by $\mathbf{A}$ preserves or reverses the orientation.

Explain how this is connected to your answer to part (ii).

Solution

(i) Matrix $\mathbf{A}$ represents a reflection in the plane $x = 0$.

(ii) Using a calculator, det $\mathbf{A} = -1$.

(iii) Matrix $\mathbf{A}$ represents a reflection, so the orientation is reversed. This is confirmed by the negative determinant.

> The first column of $\mathbf{A}$ shows that the unit vector $\mathbf{i}$ is mapped to $-\mathbf{i}$, and the other columns show that the unit vectors $\mathbf{j}$ and $\mathbf{k}$ are mapped to themselves.

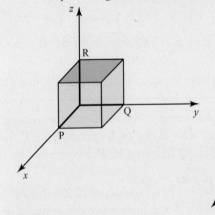
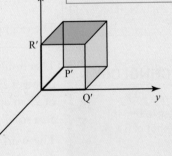

Figure 7.5

Matrices with determinant zero

Figure 7.6 shows the image of the unit square OIPJ under the transformation represented by the matrix $\mathbf{T} = \begin{pmatrix} 6 & 4 \\ 3 & 2 \end{pmatrix}$.

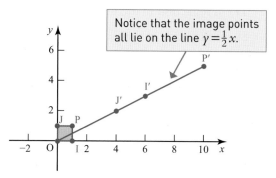

> Notice that the image points all lie on the line $y = \frac{1}{2}x$.

Figure 7.6

The determinant of $\mathbf{T} = (6 \times 2) - (4 \times 3) = 12 - 12 = 0$.

This means that the area scale factor of the transformation is zero, so any shape is transformed into a shape with area zero.

In this case, the image of a point (p, q) is given by

$$\begin{pmatrix} 6 & 4 \\ 3 & 2 \end{pmatrix} \begin{pmatrix} p \\ q \end{pmatrix} = \begin{pmatrix} 6p + 4q \\ 3p + 2q \end{pmatrix} = \begin{pmatrix} 2(3p + 2q) \\ 3p + 2q \end{pmatrix}.$$

You can see that for all the possible image points, the y-coordinate is half the x-coordinate, showing that all the image points lie on the line $y = \frac{1}{2}x$.

In this transformation, more than one point maps to the same image point.

For example, $(4, 0) \rightarrow (24, 12)$

$(0, 6) \rightarrow (24, 12)$

$(1, 4.5) \rightarrow (24, 12)$.

Discussion point

→ What is the effect of a transformation represented by a 3×3 matrix with determinant zero?

Exercise 7.1

① For each of the following matrices:
  (a) draw a diagram to show the image of the unit square under the transformation represented by the matrix
  (b) find the area of the image in part (a)
  (c) find the determinant of the matrix.

  (i) $\begin{pmatrix} 3 & -2 \\ 4 & 1 \end{pmatrix}$ (ii) $\begin{pmatrix} 4 & 0 \\ -1 & 4 \end{pmatrix}$ (iii) $\begin{pmatrix} 4 & -8 \\ 1 & -2 \end{pmatrix}$ (iv) $\begin{pmatrix} 5 & -7 \\ -3 & 2 \end{pmatrix}$

② The matrix $\begin{pmatrix} x - 3 & -3 \\ 2 & x - 5 \end{pmatrix}$ has determinant 9.

  Find the possible values of x.

③ (i) Write down the matrices **A**, **B**, **C** and **D** which represent:

A – a reflection in the x-axis

B – a reflection in the y-axis

C – a reflection in the line $y = x$

D – a reflection in the line $y = -x$

(ii) Show that each of the matrices **A**, **B**, **C** and **D** has determinant of -1.

(iii) Draw diagrams for each of the transformations **A**, **B**, **C** and **D** to demonstrate that the images of the vertices labelled anticlockwise on the unit square OIPJ are reversed to a clockwise labelling.

④ A triangle has area $6\,\text{cm}^2$. The triangle is transformed by means of the

matrix $\begin{pmatrix} 2 & 3 \\ -3 & 1 \end{pmatrix}$.

Find the area of the image of the triangle.

⑤ The two-way stretch with matrix $\begin{pmatrix} a & 0 \\ 0 & d \end{pmatrix}$ preserves the area (i.e. the area

of the image is equal to the area of the original shape).

What is the relationship connecting a and d?

⑥ Figure 7.7 shows the unit square and a transformation of the square.

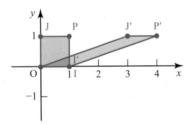

Figure 7.7

(i) Write down the matrix which represents this transformation.

(ii) Show that under this transformation the area of the image is always equal to the area of the object.

T ⑦ A transformation in three dimensions is represented by the matrix

$$\mathbf{A} = \begin{pmatrix} 2 & 3 & 1 \\ -1 & 1 & 0 \\ 0 & 4 & 2 \end{pmatrix}.$$

A cuboid has volume $5\,\text{cm}^3$. What is the volume of the image of the cuboid under the transformation represented by **A**?

⑧ $\mathbf{M} = \begin{pmatrix} 5 & 3 \\ 4 & 2 \end{pmatrix}$ and $\mathbf{N} = \begin{pmatrix} 3 & 2 \\ -2 & 1 \end{pmatrix}$.

(i) Find the determinants of **M** and **N**.

(ii) Find the matrix **MN** and show that det $(\mathbf{MN})$ = det **M** × det **N**.

⑨ The plane is transformed by the matrix $\mathbf{M} = \begin{pmatrix} 4 & -6 \\ 2 & -3 \end{pmatrix}$.

  (i) Draw a diagram to show the image of the unit square under the transformation represented by $\mathbf{M}$.

  (ii) Describe the effect of the transformation and explain this with reference to the determinant of $\mathbf{M}$.

⑩ The plane is transformed by the matrix $\mathbf{N} = \begin{pmatrix} 5 & -10 \\ -1 & 2 \end{pmatrix}$.

  (i) Find the image of the point (p, q).

  (ii) Hence show that the whole plane is mapped to a straight line and find the equation of this line.

  (iii) Find the determinant of $\mathbf{N}$ and explain its significance.

⑪ A matrix $\mathbf{T}$ maps all points on the line $x + 2y = 1$ to the point $(1, 3)$.

  (i) Find the matrix $\mathbf{T}$ and show that it has determinant of zero.

  (ii) Show that $\mathbf{T}$ maps all points on the plane to the line $y = 3x$.

  (iii) Find the coordinates of the point to which all points on the line $x + 2y = 3$ are mapped.

⑫ The plane is transformed using the matrix $\begin{pmatrix} a & b \\ c & d \end{pmatrix}$ where $ad - bc = 0$.

  Prove that the general point $P(x, y)$ maps to P' on the line $cx - ay = 0$.

⑬ The point P is mapped to P' on the line $3y = x$ so that PP' is parallel to the line $y = 3x$.

  (i) Find the equation of the line parallel to $y = 3x$ passing through the point P with coordinates (s, t).

  (ii) Find the coordinates of P', the point where this line meets $3y = x$.

  (iii) Find the matrix of the transformation which maps P to P' and show that the determinant of this matrix is zero.

2 The inverse of a matrix

The identity matrix

Whenever you multiply a 2×2 matrix $\mathbf{M}$ by $\begin{pmatrix} 1 & 0 \\ 0 & 1 \end{pmatrix}$ the product is $\mathbf{M}$.

It makes no difference whether you **pre-multiply**, for example,

$$\begin{pmatrix} 1 & 0 \\ 0 & 1 \end{pmatrix} \begin{pmatrix} 4 & -2 \\ 6 & 3 \end{pmatrix} = \begin{pmatrix} 4 & -2 \\ 6 & 3 \end{pmatrix}$$

or **post-multiply**

$$\begin{pmatrix} 4 & -2 \\ 6 & 3 \end{pmatrix} \begin{pmatrix} 1 & 0 \\ 0 & 1 \end{pmatrix} = \begin{pmatrix} 4 & -2 \\ 6 & 3 \end{pmatrix}.$$

The matrix $\begin{pmatrix} 1 & 0 \\ 0 & 1 \end{pmatrix}$ is known as the 2 × 2 identity matrix.

Identity matrices are often denoted by the letter **I**.

For multiplication of matrices, **I** behaves in the same way as the number 1 when dealing with the multiplication of real numbers.

The transformation represented by the identity matrix maps every point to itself.

Similarly, the 3 × 3 identity matrix is $\begin{pmatrix} 1 & 0 & 0 \\ 0 & 1 & 0 \\ 0 & 0 & 1 \end{pmatrix}$.

Example 7.5

(i) Write down the matrix **A** which represents a rotation of 90° anticlockwise about the origin.

(ii) Write down the matrix **B** which represents a rotation of 90° clockwise about the origin.

(iii) Find the product **AB** and comment on your answer.

Solution

(i) $\mathbf{A} = \begin{pmatrix} 0 & -1 \\ 1 & 0 \end{pmatrix}$

(ii) $\mathbf{B} = \begin{pmatrix} 0 & 1 \\ -1 & 0 \end{pmatrix}$

(iii) $\mathbf{AB} = \begin{pmatrix} 0 & -1 \\ 1 & 0 \end{pmatrix}\begin{pmatrix} 0 & 1 \\ -1 & 0 \end{pmatrix} = \begin{pmatrix} 1 & 0 \\ 0 & 1 \end{pmatrix}$

AB represents a rotation of 90° clockwise followed by a rotation of 90° anticlockwise. The result of this is to return to the starting point.

To undo the effect of a rotation through 90° anticlockwise about the origin, you need to carry out a rotation through 90° clockwise about the origin. These two transformations are inverses of each other.

Similarly, the matrices which represent these transformations are inverses of each other.

In Example 7.5, $\mathbf{B} = \begin{pmatrix} 0 & 1 \\ -1 & 0 \end{pmatrix}$ is the inverse of $\mathbf{A} = \begin{pmatrix} 0 & -1 \\ 1 & 0 \end{pmatrix}$, and vice versa.

Finding the inverse of a matrix

If the product of two square matrices, **M** and **N**, is the identity matrix **I**, then **N** is the inverse of **M**. You can write this as $\mathbf{N} = \mathbf{M}^{-1}$.

Note

The notation $\mathbf{M}^{-1}$ means the inverse of matrix **M**, not $\dfrac{1}{\mathbf{M}}$.

Generally, if $\mathbf{M} = \begin{pmatrix} a & b \\ c & d \end{pmatrix}$ you need to find an inverse matrix $\begin{pmatrix} p & q \\ r & s \end{pmatrix}$ such that $\begin{pmatrix} a & b \\ c & d \end{pmatrix}\begin{pmatrix} p & q \\ r & s \end{pmatrix} = \begin{pmatrix} 1 & 0 \\ 0 & 1 \end{pmatrix}$.

ACTIVITY 7.4

Multiply $\begin{pmatrix} a & b \\ c & d \end{pmatrix}$ by $\begin{pmatrix} d & -b \\ -c & a \end{pmatrix}$.

What do you notice?

Use your result to write down the inverse of the general matrix $\mathbf{M} = \begin{pmatrix} a & b \\ c & d \end{pmatrix}$.
How does the determinant $|\mathbf{M}|$ relate to the matrix $\mathbf{M}^{-1}$?

You should have found in the activity that the inverse of the matrix

$\mathbf{M} = \begin{pmatrix} a & b \\ c & d \end{pmatrix}$ is given by

$$\mathbf{M}^{-1} = \frac{1}{ad - bc} \begin{pmatrix} d & -b \\ -c & a \end{pmatrix}.$$

If $\det \mathbf{M} = 0$ then the inverse matrix $\mathbf{M}^{-1}$ does not exist and $\mathbf{M}$ is said to be **singular**. If $\det \mathbf{M} \neq 0$ then $\mathbf{M}$ is said to be **non-singular**.

If a matrix is singular, then it maps all points on the plane to a straight line. So an infinite number of points are mapped to the same point on the straight line. It is therefore not possible to find the inverse of the transformation, because an inverse matrix would map a point on that straight line to just one other point, not to an infinite number of them.

> A special case is the zero matrix, which maps all points to the origin.

You can use your calculator to find the inverse of a matrix with numerical entries.

You will learn how to find the inverse of a 3×3 matrix in the next section. However, you may sometimes be able to spot the inverse of a 3×3 matrix if you have already found a matrix product which has turned out to be a multiple of the identity matrix.

$$\mathbf{MN} = k\,\mathbf{I} \quad \text{so } \mathbf{M}^{-1} = \frac{1}{k}\mathbf{N}, \text{ provided } k \neq 0$$

Example 7.6

$$\mathbf{A} = \begin{pmatrix} 11 & 3 \\ 6 & 2 \end{pmatrix}$$

(i)   Find $\mathbf{A}^{-1}$.

(ii)   The point P is mapped to the point Q $(5, 2)$ under the transformation represented by $\mathbf{A}$. Find the coordinates of P.

Solution

(i) $\det \mathbf{A} = (11 \times 2) - (3 \times 6) = 4$

$$\mathbf{A}^{-1} = \frac{1}{4} \begin{pmatrix} 2 & -3 \\ -6 & 11 \end{pmatrix}$$

(ii) $\mathbf{A}^{-1}\begin{pmatrix} 5 \\ 2 \end{pmatrix} = \frac{1}{4}\begin{pmatrix} 2 & -3 \\ -6 & 11 \end{pmatrix}\begin{pmatrix} 5 \\ 2 \end{pmatrix}$

$= \frac{1}{4}\begin{pmatrix} 4 \\ -8 \end{pmatrix}$

$= \begin{pmatrix} 1 \\ -2 \end{pmatrix}$

$P = (1, -2)$

> **A** maps P to Q, so **A**$^{-1}$ maps Q to P.

As matrix multiplication is generally non-commutative, it is interesting to find out if $\mathbf{MM}^{-1} = \mathbf{M}^{-1}\mathbf{M}$. The next activity investigates this.

ACTIVITY 7.5

(i) In Example 7.6 you found that the inverse of $\mathbf{A} = \begin{pmatrix} 11 & 3 \\ 6 & 2 \end{pmatrix}$ is

$\mathbf{A}^{-1} = \frac{1}{4}\begin{pmatrix} 2 & -3 \\ -6 & 11 \end{pmatrix}$.

Show that $\mathbf{AA}^{-1} = \mathbf{A}^{-1}\mathbf{A} = \mathbf{I}$.

(ii) If the matrix $\mathbf{M} = \begin{pmatrix} a & b \\ c & d \end{pmatrix}$, write down $\mathbf{M}^{-1}$ and show that $\mathbf{MM}^{-1} = \mathbf{M}^{-1}\mathbf{M} = \mathbf{I}$.

Discussion points

➔ How would you reverse the effect of a rotation followed by a reflection?

➔ How would you write down the inverse of a matrix product **MN** in terms of $\mathbf{M}^{-1}$ and $\mathbf{N}^{-1}$?

The result $\mathbf{MM}^{-1} = \mathbf{M}^{-1}\mathbf{M} = \mathbf{I}$ is important as it means that the inverse of a matrix, if it exists, is unique. This is true for all square matrices, not just 2×2 matrices.

The inverse of a product of matrices

Suppose you want to find the inverse of the product **MN**, where **M** and **N** are non-singular matrices. This means that you need to find a matrix **X** such that

$\mathbf{X}(\mathbf{MN}) = \mathbf{I}$.

$\mathbf{X}(\mathbf{MN}) = \mathbf{I} \Rightarrow \mathbf{XMNN}^{-1} = \mathbf{IN}^{-1}$ ← Post multiply by $\mathbf{N}^{-1}$

$\Rightarrow \mathbf{XM} = \mathbf{IN}^{-1}$ ← Using $\mathbf{NN}^{-1} = \mathbf{I}$

$\Rightarrow \mathbf{XMM}^{-1} = \mathbf{N}^{-1}\mathbf{M}^{-1}$ ← Post multiply by $\mathbf{M}^{-1}$

$\Rightarrow \mathbf{X} = \mathbf{N}^{-1}\mathbf{M}^{-1}$ ← Using $\mathbf{MM}^{-1} = \mathbf{I}$

So $(\mathbf{MN})^{-1} = \mathbf{N}^{-1}\mathbf{M}^{-1}$ for matrices **M** and **N** of the same order. This means that when working backwards, you must reverse the second transformation before reversing the first transformation.

⌨ TECHNOLOGY

Investigate how to use your calculator to find the inverse of 2×2 and 3×3 matrices.

Check using your calculator that multiplying a matrix by its inverse gives the identity matrix.

Exercise 7.2

① For the matrix $\begin{pmatrix} 5 & -1 \\ -2 & 0 \end{pmatrix}$

(i)   find the image of the point $(3, 5)$

(ii) find the inverse matrix

(iii) find the point which maps to the image $(3, -2)$.

② Determine whether the following matrices are singular or non-singular. For those that are non-singular, find the inverse.

(i) $\begin{pmatrix} 6 & 3 \\ -4 & 2 \end{pmatrix}$ (ii) $\begin{pmatrix} 6 & 3 \\ 4 & 2 \end{pmatrix}$ (iii) $\begin{pmatrix} 11 & 3 \\ 3 & 11 \end{pmatrix}$ (iv) $\begin{pmatrix} 11 & 11 \\ 3 & 3 \end{pmatrix}$

(v) $\begin{pmatrix} 2 & -7 \\ 0 & 0 \end{pmatrix}$ (vi) $\begin{pmatrix} -2a & 4a \\ 4b & -8b \end{pmatrix}$ (vii) $\begin{pmatrix} -2 & 4a \\ 4b & -8 \end{pmatrix}$

③ Using a calculator, find whether the following matrices are singular or non-singular. For those that are non-singular find the inverse.

(i) $\begin{pmatrix} 2 & 4 & 9 \\ -1 & -3 & 0 \\ 4 & -2 & -7 \end{pmatrix}$ (ii) $\begin{pmatrix} 4 & 0 & -1 \\ 2 & -3 & 5 \\ -4 & 6 & -10 \end{pmatrix}$ (iii) $\begin{pmatrix} 1 & 0 & 3 \\ 8 & -2 & -1 \\ 3 & 5 & 11 \end{pmatrix}$

④ $\mathbf{M} = \begin{pmatrix} 5 & 6 \\ 2 & 3 \end{pmatrix}$ and $\mathbf{N} = \begin{pmatrix} 8 & 5 \\ -2 & -1 \end{pmatrix}$.

Calculate the following:

(i) $\mathbf{M}^{-1}$ (iii) $\mathbf{MN}$ (v) $(\mathbf{MN})^{-1}$ (vii) $\mathbf{M}^{-1}\mathbf{N}^{-1}$

(ii) $\mathbf{N}^{-1}$ (iv) $\mathbf{NM}$ (vi) $(\mathbf{NM})^{-1}$ (viii) $\mathbf{N}^{-1}\mathbf{M}^{-1}$

⑤ The diagram shows the unit square OIPJ mapped to the image OI′P′J′ under a transformation represented by a matrix $\mathbf{M}$.

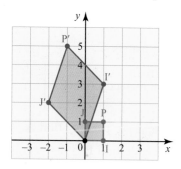

Figure 7.8

(i) Find the inverse of $\mathbf{M}$.

(ii) Use matrix multiplication to show that $\mathbf{M}^{-1}$ maps OI′P′J′ back to OIPJ.

⑥ The matrix $\begin{pmatrix} 1-k & 2 \\ -1 & 4-k \end{pmatrix}$ is singular.

Find the possible values of k.

⑦ Given that $\mathbf{M} = \begin{pmatrix} 2 & 3 \\ -1 & 4 \end{pmatrix}$ and $\mathbf{MN} = \begin{pmatrix} 7 & 2 & -9 & 10 \\ 2 & -1 & -12 & 17 \end{pmatrix}$, find the matrix $\mathbf{N}$.

⑧ Triangle T has vertices at $(1, 0)$, $(0, 1)$ and $(-2, 0)$.

It is transformed to triangle T′ by the matrix $\mathbf{M} = \begin{pmatrix} 3 & 1 \\ 1 & 1 \end{pmatrix}$.

(i) Find the coordinates of the vertices of T′.

Show the triangles T and T′ on a single diagram.

(ii) Find the ratio of the area of T′ to the area of T.

Comment on your answer in relation to the matrix $\mathbf{M}$.

(iii) Find $\mathbf{M}^{-1}$ and verify that this matrix maps the vertices of T′ to the vertices of T.

⑨ $\mathbf{M} = \begin{pmatrix} a & b \\ c & d \end{pmatrix}$ is a singular matrix.

(i) Show that $\mathbf{M}^2 = (a + d)\mathbf{M}$.

(ii) Find a formula which expresses $\mathbf{M}^n$ in terms of $\mathbf{M}$, where n is a positive integer.

Comment on your results.

⑩ Given that $\mathbf{PQR} = \mathbf{I}$, show algebraically that

(i) $\mathbf{Q} = \mathbf{P}^{-1}\mathbf{R}^{-1}$

(ii) $\mathbf{Q}^{-1} = \mathbf{RP}$.

Given that $\mathbf{P} = \begin{pmatrix} 3 & 1 \\ 1 & 2 \end{pmatrix}$ and $\mathbf{R} = \begin{pmatrix} 12 & -3 \\ 2 & -1 \end{pmatrix}$

(iii) use part (i) to find the matrix $\mathbf{Q}$

(iv) calculate the matrix $\mathbf{Q}^{-1}$

(v) verify that your answer to part (ii) is correct by calculating $\mathbf{RP}$ and comparing it with your answer to part (iv).

⑪ $\mathbf{A} = \begin{pmatrix} 1 & 7 & 4 \\ 0 & 1 & 2 \\ 0 & 0 & 1 \end{pmatrix}$, $\mathbf{B} = \begin{pmatrix} 1 & 0 & 0 \\ 3 & 1 & 0 \\ -1 & -4 & 1 \end{pmatrix}$ and $\mathbf{C} = \mathbf{AB}$.

(i) Calculate the matrix $\mathbf{C}$.

(ii) Work out the matrix product $\mathbf{A}\begin{pmatrix} 1 & a & b \\ 0 & 1 & c \\ 0 & 0 & 1 \end{pmatrix}$.

(iii) Using the answer to part (ii), find $\mathbf{A}^{-1}$.

(iv) Using a calculator, find $\mathbf{B}^{-1}$.

(v) Using your results from parts (iii) and (iv), find $\mathbf{C}^{-1}$.

⑫ The matrix $\mathbf{M} = \begin{pmatrix} k-1 & k-1 & 0 \\ 1 & k+1 & -2 \\ k-1 & k-2 & 1 \end{pmatrix}$ has inverse

$$\mathbf{M}^{-1} = \begin{pmatrix} k & -1 & -2 \\ -\dfrac{5}{2} & k-2 & k-1 \\ -\dfrac{7}{2} & 1 & k \end{pmatrix}.$$

Find the value of k.

3 The determinant and inverse of a 3 × 3 matrix

You have seen that you can use a calculator to find the determinant and inverse of a 3×3 matrix.

It is also possible to find the determinant and inverse of a 3×3 matrix without using a calculator. This is useful in cases where some of the elements of the matrix are algebraic rather than numerical.

If $\mathbf{M}$ is the 3×3 matrix $\begin{pmatrix} a_1 & b_1 & c_1 \\ a_2 & b_2 & c_2 \\ a_3 & b_3 & c_3 \end{pmatrix}$ then the determinant of $\mathbf{M}$ is defined by

$$\det \mathbf{M} = a_1 \begin{vmatrix} b_2 & c_2 \\ b_3 & c_3 \end{vmatrix} - a_2 \begin{vmatrix} b_1 & c_1 \\ b_3 & c_3 \end{vmatrix} + a_3 \begin{vmatrix} b_1 & c_1 \\ b_2 & c_2 \end{vmatrix},$$

which is sometimes referred to as the **expansion of the determinant by the first column**.

For example, to find the determinant of the matrix $\mathbf{A} = \begin{pmatrix} 3 & -2 & 1 \\ 0 & 1 & 2 \\ 4 & 0 & 1 \end{pmatrix}$:

> Notice that you do not really need to calculate $\begin{vmatrix} -2 & 1 \\ 0 & 1 \end{vmatrix}$ as it is going to be multiplied by zero. Keeping an eye open for helpful zeros can reduce the number of calculations needed.

$$\det \mathbf{A} = 3 \begin{vmatrix} 1 & 2 \\ 0 & 1 \end{vmatrix} - 0 \begin{vmatrix} -2 & 1 \\ 0 & 1 \end{vmatrix} + 4 \begin{vmatrix} -2 & 1 \\ 1 & 2 \end{vmatrix}$$

$$= 3(1 - 0) - 0(-2 - 0) + 4(-4 - 1)$$

$$= 3 - 20$$

$$= -17$$

The 2×2 determinant $\begin{vmatrix} b_2 & c_2 \\ b_3 & c_3 \end{vmatrix}$ is called the **minor** of the element a_1.

It is obtained by deleting the row and column containing a_1:

$$\begin{vmatrix} \cancel{a_1} & \cancel{b_1} & \cancel{c_1} \\ \cancel{a_2} & b_2 & c_2 \\ \cancel{a_3} & b_3 & c_3 \end{vmatrix}$$

Other minors are defined in the same way, for example the minor of a_2 is

$$\begin{vmatrix} \cancel{a_1} & b_1 & c_1 \\ \cancel{a_2} & \cancel{b_2} & \cancel{c_2} \\ \cancel{a_3} & b_3 & c_3 \end{vmatrix} = \begin{vmatrix} b_1 & c_1 \\ b_3 & c_3 \end{vmatrix}$$

> **Note**
>
> As an alternative, you could use the **expansion of the determinant by the second column:**
>
> $$\det \mathbf{M} = -b_1 \begin{vmatrix} a_2 & c_2 \\ a_3 & c_3 \end{vmatrix} + b_2 \begin{vmatrix} a_1 & c_1 \\ a_3 & c_3 \end{vmatrix} - b_3 \begin{vmatrix} a_1 & c_1 \\ a_2 & c_2 \end{vmatrix},$$
>
> or the **expansion of the determinant by the third column:**
>
> $$\det \mathbf{M} = c_1 \begin{vmatrix} a_2 & b_2 \\ a_3 & b_3 \end{vmatrix} - c_2 \begin{vmatrix} a_1 & b_1 \\ a_3 & b_3 \end{vmatrix} + c_3 \begin{vmatrix} a_1 & b_1 \\ a_2 & b_2 \end{vmatrix}.$$
>
> It is fairly easy to show that all three expressions above for $\det \mathbf{M}$ simplify to:
>
> $$a_1 b_2 c_3 + a_2 b_3 c_1 + a_3 b_1 c_2 - a_3 b_2 c_1 - a_1 b_3 c_2 - a_2 b_1 c_3$$

You may have noticed that the signs on the minors alternate as shown:

$$\begin{vmatrix} + & - & + \\ - & + & - \\ + & - & + \end{vmatrix}$$

A minor, together with its correct sign, is known as a **cofactor** and is denoted by the corresponding capital letter; for example, the cofactor of a_3 is A_3. This means that the expansion by the first column, say, can be written as

$$a_1 A_1 + a_2 A_2 + a_3 A_3.$$

Example 7.7

Find the determinant of the matrix $\mathbf{M} = \begin{pmatrix} 3 & 0 & -4 \\ 7 & 2 & -1 \\ -2 & 1 & 3 \end{pmatrix}$

Solution

Expanding by the first column using the expression:

$$\det \mathbf{M} = a_1 \begin{vmatrix} b_2 & c_2 \\ b_3 & c_3 \end{vmatrix} - a_2 \begin{vmatrix} b_1 & c_1 \\ b_3 & c_3 \end{vmatrix} + a_3 \begin{vmatrix} b_1 & c_1 \\ b_2 & c_2 \end{vmatrix}$$

gives:

> To find the determinant you can also expand by rows. So, for example, expanding by the top row would give:
>
> $$3 \begin{vmatrix} 2 & -1 \\ 1 & 3 \end{vmatrix} - 0 \begin{vmatrix} 7 & -1 \\ -2 & 3 \end{vmatrix}$$
> $$+ (-4) \begin{vmatrix} 7 & 2 \\ -2 & 1 \end{vmatrix}$$
>
> which also gives the answer −23.

$$\det \mathbf{M} = 3 \begin{vmatrix} 2 & -1 \\ 1 & 3 \end{vmatrix} - 7 \begin{vmatrix} 0 & -4 \\ 1 & 3 \end{vmatrix} + (-2) \begin{vmatrix} 0 & -4 \\ 2 & -1 \end{vmatrix}$$

$$= 3(6 - (-1)) - 7(0 - (-4)) - 2(0 - (-8))$$

$$= 21 - 28 - 16$$

$$= -23$$

> Notice that expanding by the top row would be quicker here as it has a zero element.

Earlier you saw that the determinant of a 2 × 2 matrix represents the area scale factor of the transformation represented by the matrix. In the case of a 3 × 3 matrix the determinant represents the volume scale factor. For example, the matrix $\begin{pmatrix} 2 & 0 & 0 \\ 0 & 2 & 0 \\ 0 & 0 & 2 \end{pmatrix}$ has determinant 8; this matrix represents an enlargement of scale factor 2, centre the origin, so the volume scale factor of the transformation is $2 \times 2 \times 2 = 8$.

The matrix $\begin{pmatrix} A_1 & A_2 & A_3 \\ B_1 & B_2 & B_3 \\ C_1 & C_2 & C_3 \end{pmatrix}$ is known as the **adjugate** or **adjoint** of **M**, denoted **adj M**.

> Recall that a minor, together with its correct sign, is known as a cofactor and is denoted by the corresponding capital letter, for example, the cofactor of a_3 is A_3.

The adjugate of **M** is formed by

- replacing each element of **M** by its cofactor
- then transposing the matrix (i.e. changing rows into columns and columns into rows).

The unique inverse of a 3 × 3 matrix can be calculated as follows:

$$\mathbf{M}^{-1} = \frac{1}{\det \mathbf{M}} \text{adj } \mathbf{M} = \frac{1}{\det \mathbf{M}} \begin{pmatrix} A_1 & A_2 & A_3 \\ B_1 & B_2 & B_3 \\ C_1 & C_2 & C_3 \end{pmatrix}, \det \mathbf{M} \neq 0$$

The steps involved in the method are shown in the following example.

Example 7.8

Find the inverse of the matrix **M** without using a calculator, where

$$\mathbf{M} = \begin{pmatrix} 2 & 3 & 4 \\ 2 & -5 & 2 \\ -3 & 6 & -3 \end{pmatrix}.$$

Solution

Step 1: Find the determinant Δ and check $\Delta \neq 0$.

Expanding by the first column

$$\Delta = 2 \begin{vmatrix} -5 & 2 \\ 6 & -3 \end{vmatrix} - 2 \begin{vmatrix} 3 & 4 \\ 6 & -3 \end{vmatrix} - 3 \begin{vmatrix} 3 & 4 \\ -5 & 2 \end{vmatrix}$$

$$= (2 \times 3) - (2 \times -33) - (3 \times 26) = -6$$

Therefore the inverse matrix exists.

Step 2: Evaluate the cofactors.

$$A_1 = \begin{vmatrix} -5 & 2 \\ 6 & -3 \end{vmatrix} = 3 \quad A_2 = -\begin{vmatrix} 3 & 4 \\ 6 & -3 \end{vmatrix} = 33 \quad A_3 = \begin{vmatrix} 3 & 4 \\ -5 & 2 \end{vmatrix} = 26$$

$$B_1 = -\begin{vmatrix} 2 & 2 \\ -3 & -3 \end{vmatrix} = 0 \qquad B_2 = \begin{vmatrix} 2 & 4 \\ -3 & -3 \end{vmatrix} = 6 \qquad B_3 = -\begin{vmatrix} 2 & 4 \\ 2 & 2 \end{vmatrix} = 4$$

$$C_1 = \begin{vmatrix} 2 & -5 \\ -3 & 6 \end{vmatrix} = -3 \qquad C_2 = -\begin{vmatrix} 2 & 3 \\ -3 & 6 \end{vmatrix} = -21 \qquad C_3 = \begin{vmatrix} 2 & 3 \\ 2 & -5 \end{vmatrix} = -16$$

> You can evaluate the determinant Δ using these cofactors to check your earlier arithmetic is correct:
>
> 2nd column: $\Delta = 3B_1 - 5B_2 + 6B_3$
> $$= (3 \times 0) - (5 \times 6) + (6 \times 4) = -6$$
>
> 3rd column: $\Delta = 4C_1 + 2C_2 - 3C_3$
> $$= (4 \times -3) + (2 \times -21) - (3 \times -16) = -6$$

Step 3: Form the matrix of cofactors and transpose it, then multiply by $\dfrac{1}{\Delta}$.

$$\mathbf{M}^{-1} = \frac{1}{-6} \begin{pmatrix} 3 & 0 & -3 \\ 33 & 6 & -21 \\ 26 & 4 & -16 \end{pmatrix}^T$$

The capital T indicates the matrix is to be transposed.

Matrix of cofactors.

Multiply by $\dfrac{1}{\Delta}$.

$$= \frac{1}{-6} \begin{pmatrix} 3 & 33 & 26 \\ 0 & 6 & 4 \\ -3 & -21 & -16 \end{pmatrix}$$

$$= \frac{1}{6} \begin{pmatrix} -3 & -33 & -26 \\ 0 & -6 & -4 \\ 3 & 21 & 16 \end{pmatrix}$$

The final matrix could then be simplified and written as

$$\mathbf{M}^{-1} = \begin{pmatrix} -\dfrac{1}{2} & -\dfrac{11}{2} & -\dfrac{13}{3} \\[2mm] 0 & -1 & -\dfrac{2}{3} \\[2mm] \dfrac{1}{2} & \dfrac{7}{2} & \dfrac{8}{3} \end{pmatrix}$$

Check: $\mathbf{M}\mathbf{M}^{-1} = \begin{pmatrix} 2 & 3 & 4 \\ 2 & -5 & 2 \\ -3 & 6 & -3 \end{pmatrix} \dfrac{1}{6} \begin{pmatrix} -3 & -33 & -26 \\ 0 & -6 & -4 \\ 3 & 21 & 16 \end{pmatrix}$

$$= \frac{1}{6} \begin{pmatrix} 6 & 0 & 0 \\ 0 & 6 & 0 \\ 0 & 0 & 6 \end{pmatrix} = \begin{pmatrix} 1 & 0 & 0 \\ 0 & 1 & 0 \\ 0 & 0 & 1 \end{pmatrix}$$

This adjugate method for finding the inverse of a 3 × 3 matrix is reasonably straightforward but it is important to check your arithmetic as you go along, as it is very easy to make mistakes. You can use your calculator to check that you have calculated the inverse correctly.

As shown in the example above, you might also multiply the inverse by the original matrix and check that you obtain the 3 × 3 identity matrix.

Exercise 7.3

① Evaluate these determinants without using a calculator. Check your answers using your calculator.

(i) (a) $\begin{vmatrix} 1 & 1 & 3 \\ -1 & 0 & 2 \\ 3 & 1 & 4 \end{vmatrix}$
(b) $\begin{vmatrix} 1 & -1 & 3 \\ 1 & 0 & 1 \\ 3 & 2 & 4 \end{vmatrix}$

(ii) (a) $\begin{vmatrix} 1 & -5 & -4 \\ 2 & 3 & 3 \\ -2 & 1 & 0 \end{vmatrix}$
(b) $\begin{vmatrix} 1 & 2 & -2 \\ -5 & 3 & 1 \\ -4 & 3 & 0 \end{vmatrix}$

(iii) (a) $\begin{vmatrix} 2 & 1 & 2 \\ 3 & 5 & 3 \\ 1 & -1 & 1 \end{vmatrix}$
(b) $\begin{vmatrix} 1 & 5 & 0 \\ 1 & 5 & 0 \\ 2 & 1 & -2 \end{vmatrix}$

What do you notice about the determinants?

② Find the inverses of the following matrices, if they exist, without using a calculator.

(i) $\begin{pmatrix} 1 & 2 & 4 \\ 2 & 4 & 5 \\ 0 & 1 & 2 \end{pmatrix}$
(ii) $\begin{pmatrix} 3 & 2 & 6 \\ 5 & 3 & 11 \\ 7 & 4 & 16 \end{pmatrix}$

(iii) $\begin{pmatrix} 5 & 5 & -5 \\ -9 & 3 & -5 \\ -4 & -6 & 8 \end{pmatrix}$
(iv) $\begin{pmatrix} 6 & 5 & 6 \\ -5 & 2 & -4 \\ -4 & -6 & -5 \end{pmatrix}$

③ Find the inverse of the matrix $\mathbf{M} = \begin{pmatrix} 1 & 3 & -2 \\ k & 0 & 4 \\ 2 & -1 & 4 \end{pmatrix}$ where $k \neq 0$.

For what value of k is the matrix $\mathbf{M}$ singular?

④ (i) Investigate the relationship between the matrices

$$\mathbf{A} = \begin{pmatrix} 0 & 3 & 1 \\ 2 & 4 & 2 \\ -1 & 3 & 5 \end{pmatrix} \quad \mathbf{B} = \begin{pmatrix} 1 & 0 & 3 \\ 2 & 2 & 4 \\ 5 & -1 & 3 \end{pmatrix} \quad \mathbf{C} = \begin{pmatrix} 3 & 1 & 0 \\ 4 & 2 & 2 \\ 3 & 5 & -1 \end{pmatrix}$$

(ii) Find det $\mathbf{A}$, det $\mathbf{B}$ and det $\mathbf{C}$ and comment on your answer.

⑤ Show that $x = 1$ is one root of the equation $\begin{vmatrix} 2 & 2 & x \\ 1 & x & 1 \\ x & 1 & 4 \end{vmatrix} = 0$ and find the other roots.

⑥ Find the values of x for which the matrix $\begin{pmatrix} 3 & -1 & 1 \\ 2 & x & 4 \\ x & 1 & 3 \end{pmatrix}$ is singular.

⑦ Given that the matrix $\mathbf{M} = \begin{pmatrix} k & 2 & 1 \\ 0 & -k & 2 \\ 2k & 1 & 3 \end{pmatrix}$ has determinant greater than 5, find the range of possible values for k.

⑧ (i) $\mathbf{P}$ and $\mathbf{Q}$ are non-singular matrices. Prove that $\left(\mathbf{PQ}\right)^{-1} = \mathbf{Q}^{-1}\mathbf{P}^{-1}$

(ii) Find the inverses of the matrices $\mathbf{P} = \begin{pmatrix} 0 & 3 & -1 \\ -2 & 2 & 2 \\ -3 & 0 & 1 \end{pmatrix}$ and

$\mathbf{Q} = \begin{pmatrix} 2 & 1 & 2 \\ 1 & 0 & 1 \\ 4 & -3 & 2 \end{pmatrix}$.

Using the result from part (i), find $\left(\mathbf{PQ}\right)^{-1}$

⑨ (i) Prove that $\begin{vmatrix} ka_1 & b_1 & c_1 \\ ka_2 & b_2 & c_2 \\ ka_3 & b_3 & c_3 \end{vmatrix} = k \begin{vmatrix} a_1 & b_1 & c_1 \\ a_2 & b_2 & c_2 \\ a_3 & b_3 & c_3 \end{vmatrix}$, where k is a constant.

(ii) Explain in terms of volumes why multiplying all the elements in the first column by a constant k multiplies the value of the determinant by k.

(iii) What happens if you multiply a different column by k?

⑩ Given that $\begin{vmatrix} 1 & 2 & 3 \\ 6 & 4 & 5 \\ 7 & 5 & 1 \end{vmatrix} = 43$, write down the values of the determinants of:

(i) $\begin{vmatrix} 10 & 2 & 3 \\ 60 & 4 & 5 \\ 70 & 5 & 1 \end{vmatrix}$

(ii) $\begin{vmatrix} 4 & 10 & -21 \\ 24 & 20 & -35 \\ 28 & 25 & -7 \end{vmatrix}$

(iii) $\begin{vmatrix} x & 4 & 3y \\ 6x & 8 & 5y \\ 7x & 10 & y \end{vmatrix}$

(iv) $\begin{vmatrix} x^4 & \dfrac{1}{x} & 12y \\ 6x^4 & \dfrac{2}{x} & 20y \\ 7x^4 & \dfrac{5}{2x} & 4y \end{vmatrix}$

4 Using matrices to solve simultaneous equations

There are a number of methods to solve a pair of linear simultaneous equations of the form

$$3x + 2y = 17$$
$$2x - 5y = 24$$

These include elimination, substitution or graphical methods.

An alternative method involves the use of inverse matrices. This method has the advantage that it can more easily be extended to solving a set of n equations in n variables.

Example 7.9

Use a matrix method to solve the simultaneous equations

$$3x + 2y = 17$$
$$2x - 5y = 24$$

Solution

$$\begin{pmatrix} 3 & 2 \\ 2 & -5 \end{pmatrix} \begin{pmatrix} x \\ y \end{pmatrix} = \begin{pmatrix} 17 \\ 24 \end{pmatrix}.$$

Write the equations in matrix form.

The inverse of the matrix $\begin{pmatrix} 3 & 2 \\ 2 & -5 \end{pmatrix}$ is $-\dfrac{1}{19}\begin{pmatrix} -5 & -2 \\ -2 & 3 \end{pmatrix}$.

$$-\frac{1}{19}\begin{pmatrix} -5 & -2 \\ -2 & 3 \end{pmatrix}\begin{pmatrix} 3 & 2 \\ 2 & -5 \end{pmatrix}\begin{pmatrix} x \\ y \end{pmatrix} = -\frac{1}{19}\begin{pmatrix} -5 & -2 \\ -2 & 3 \end{pmatrix}\begin{pmatrix} 17 \\ 24 \end{pmatrix}$$

Pre-multiply both sides of the matrix equation by the inverse matrix.

$$\begin{pmatrix} x \\ y \end{pmatrix} = -\frac{1}{19}\begin{pmatrix} -133 \\ 38 \end{pmatrix} = \begin{pmatrix} 7 \\ -2 \end{pmatrix}$$

As $\mathbf{M}^{-1}\mathbf{M}\mathbf{p} = \mathbf{p}$ the left-hand side simplifies to $\begin{pmatrix} x \\ y \end{pmatrix}$.

The solution is $x = 7$, $y = -2$.

Similarly, three simultaneous equations in three unknowns can be represented in matrix form using a 3×3 matrix.

Note

The equations must be linear, this means they are of the form $ax + by = c$ for constants a, b and c.

Geometrical interpretation in two dimensions

Two equations in two unknowns can be represented in a plane by two straight lines. The number of points of intersection of the lines determines the number of solutions to the equations.

There are three different possibilities.

Case 1

Example 7.9 shows that two simultaneous equations can have a unique solution. Graphically, this is represented by a single point of intersection.

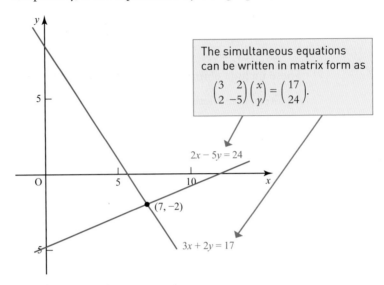

Figure 7.9

This is the case where det $\mathbf{M} \neq 0$ and so the inverse matrix $\mathbf{M}^{-1}$ exists, allowing the equations to be solved.

Case 2

If two lines are parallel they do not have a point of intersection. For example, the lines

$x + 2y = 10$

$x + 2y = 4$

are parallel.

The equations can be written in matrix form as

$$\begin{pmatrix} 1 & 2 \\ 1 & 2 \end{pmatrix} \begin{pmatrix} x \\ y \end{pmatrix} = \begin{pmatrix} 10 \\ 4 \end{pmatrix}.$$

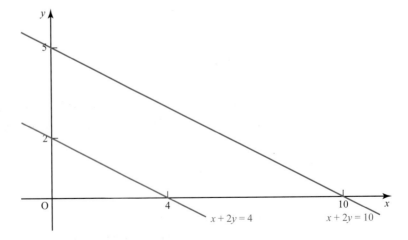

Figure 7.10

The matrix $\mathbf{M} = \begin{pmatrix} 1 & 2 \\ 1 & 2 \end{pmatrix}$ has determinant zero and hence the inverse matrix does not exist.

Case 3

More than one solution is possible in cases where the lines are **coincident**, i.e. lie on top of each other. For example, the two lines

$x + 2y = 10$
$3x + 6y = 30$

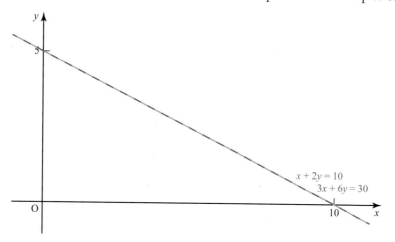

The equations can be written in matrix form as

$$\begin{pmatrix} 1 & 2 \\ 3 & 6 \end{pmatrix} \begin{pmatrix} x \\ y \end{pmatrix} = \begin{pmatrix} 10 \\ 30 \end{pmatrix}.$$

are coincident. You can see this because the equations are multiples of each other.

Figure 7.11

In this case the matrix **M** is $\begin{pmatrix} 1 & 2 \\ 3 & 6 \end{pmatrix}$ and det **M** $= 0$.

There are infinitely many solutions to these equations.

ACTIVITY 7.6

(i) Write the three simultaneous equations:

$2x - 2y + 3z = 4$
$5x + y - z = -6$
$3x + 4y - 2z = 1$

as a matrix equation.

Use a matrix method and your calculator to solve the simultaneous equations.

(ii) Repeat part (i) for the three simultaneous equations:

$2x - 2y + 3z = 4$
$5x + y - z = -6$
$3x + 3y - 4z = 1$

What happens in this case?

Try to solve the equations algebraically. Comment on your answer.

① (i) Find the inverse of the matrix $\begin{pmatrix} 3 & -1 \\ 2 & 3 \end{pmatrix}$.

(ii) Hence use a matrix method to solve the simultaneous equations

$3x - y = 2$

$2x + 3y = 5$

② Use matrices to solve the following pairs of simultaneous equations.

(i) $3x + 2y = 4$

$x - 2y = 4$

(ii) $3x - 2y = 9$

$x - 4y = -2$

③ (i) Use a calculator to find the inverse of the matrix $\begin{pmatrix} 3 & 1 & 1 \\ -5 & -2 & 3 \\ 1 & 1 & 1 \end{pmatrix}$.

(ii) Hence use a matrix method to solve the simultaneous equations

$3x + y + z = -2$

$-5x - 2y + 3z = -1$

$x + y + z = 2$

④ Use a matrix method to solve these simultaneous equations.

$x + 5y + z = 0$

$2x - 3y - 4z = 7$

$3x + 2y - 6z = 4$

⑤ For each of the following pair of equations, describe the intersections of the pair of straight lines represented by the simultaneous equations.

(i) $3x + 5y = 18$

$2x + 4y = 11$

(ii) $3x + 6y = 18$

$2x + 4y = 12$

(iii) $3x + 6y = 18$

$2x + 4y = 15$

⑥ Find the two values of k for which the equations

$2x + ky = 3$

$kx + 8y = 6$

do not have a unique solution.

How many solutions are there in each case?

⑦ (i) Find **AB** where $\mathbf{A} = \begin{pmatrix} 5 & -2 & k \\ 3 & -4 & -5 \\ -2 & 3 & 4 \end{pmatrix}$

and $\mathbf{B} = \begin{pmatrix} -1 & 3k+8 & 4k+10 \\ -2 & 2k+20 & 3k+25 \\ 1 & -11 & -14 \end{pmatrix}$.

Hence write down the inverse matrix $\mathbf{A}^{-1}$, stating the condition on the value of k required for the inverse to exist.

(ii) Using the result from part (i) solve the equation

$$\begin{pmatrix} 5 & -2 & k \\ 3 & -4 & -5 \\ -2 & 3 & 4 \end{pmatrix}\begin{pmatrix} x \\ y \\ z \end{pmatrix} = \begin{pmatrix} 28 \\ 0 \\ m \end{pmatrix}$$

when $k = 8$ and $m = 2$.

⑧ Find the conditions on a and b for which the simultaneous equations
$ax + by = 1$

$bx + ay = b$

have a unique solution.

Solve the equations when $a = -3$ giving your answers in terms of b.

Find the value of b for which the solution will lie on the line $y = -x$.

LEARNING OUTCOMES

When you have completed this chapter you should be able to:

➤ find the determinant of a 2×2 matrix

➤ know what is meant by a singular matrix

➤ understand that the determinant of a 2×2 matrix represents the area scale factor of the corresponding transformation, and understand the significance of the sign of the determinant

➤ find the inverse of a non-singular 2×2 matrix

➤ find the determinant and inverse of a 3×3 matrix, with or without a calculator

➤ know that the determinant of a 3×3 matrix represents the volume scale factor of the corresponding transformation

➤ understand the significance of a zero determinant in terms of transformations

➤ use the product rule for inverse matrices

➤ use matrices to solve a pair of linear simultaneous equations in two unknowns

➤ use matrices to solve three linear simultaneous equations in three unknowns.

KEY POINTS

1 If $\mathbf{M} = \begin{pmatrix} a & b \\ c & d \end{pmatrix}$ then the determinant of $\mathbf{M}$ is det $\mathbf{M}$, $|\mathbf{M}|$,

det $\mathbf{M}$ or $\Delta = ad - bc$.

2 If $\mathbf{M} = \begin{pmatrix} a & b \\ c & d \end{pmatrix}$ has determinant $\Delta \neq 0$, then $\mathbf{M}^{-1} = \dfrac{1}{\Delta}\begin{pmatrix} d & -b \\ -c & a \end{pmatrix}$.

3 If $\mathbf{MN} = k\,\mathbf{I}$ and $k \neq 0$ then $\mathbf{M}^{-1} = \dfrac{1}{k}\,\mathbf{N}$.

4 You should be also able to find determinants and inverses using your calculator.

5 $(\mathbf{MN})^{-1} = \mathbf{N}^{-1}\mathbf{M}^{-1}$

6 A matrix is singular if the determinant is zero. If the determinant is non-zero the matrix is said to be non-singular.

7 If the determinant of a matrix is zero, all points are mapped to either a straight line (in two dimensions) or to a plane (three dimensions)

8 If $\mathbf{A}$ is a non-singular matrix, $\mathbf{AA}^{-1} = \mathbf{A}^{-1}\mathbf{A} = \mathbf{I}$.

9 When solving two simultaneous equations in two unknowns, the equations can be written as a matrix

equation $\mathbf{M}\begin{pmatrix} x \\ y \end{pmatrix} = \begin{pmatrix} a \\ b \end{pmatrix}$.

When solving three simultaneous equations in three unknowns, the equations can be written as a matrix

equation $\mathbf{M}\begin{pmatrix} x \\ y \\ z \end{pmatrix} = \begin{pmatrix} a \\ b \\ c \end{pmatrix}$.

In both cases, if det $\mathbf{M} \neq 0$ there is a unique solution to the equations which can be found by pre-multiplying both sides of the equation by the inverse matrix $\mathbf{M}^{-1}$.

If det $\mathbf{M} = 0$ there is no unique solution to the equations. In this case there is either no solution or an infinite number of solutions.

10 The determinant of a 2×2 matrix represents the area scale factor and the determinant of a 3×3 matrix represents the volume scale factor. If the determinant is negative the orientation has changed.

11 A matrix is *singular* if its determinant is zero. If the determinant is non-zero, the matrix is said to be non-singular.

12 Two simultaneous linear equations in two unknowns,

$ax + by = p$ and $cx + dy = q$

can be written as a matrix equation

$$\begin{pmatrix} a & b \\ c & d \end{pmatrix}\begin{pmatrix} x \\ y \end{pmatrix} = \begin{pmatrix} p \\ q \end{pmatrix} \text{ or } \mathbf{M}\begin{pmatrix} x \\ y \end{pmatrix} = \begin{pmatrix} p \\ q \end{pmatrix}.$$

Similarly three simultaneous linear equations in three unknowns,

$a_1x + b_1y + c_1z = p$, $a_2x + b_2y + c_2z = q$ and $a_3x + b_3y + c_3z = r$

can be written as $\begin{pmatrix} a_1 & b_1 & c_1 \\ a_2 & b_2 & c_2 \\ a_3 & b_3 & c_3 \end{pmatrix}\begin{pmatrix} x \\ y \\ z \end{pmatrix} = \begin{pmatrix} p \\ q \\ r \end{pmatrix} \text{ or } \mathbf{M}\begin{pmatrix} x \\ y \\ z \end{pmatrix} = \begin{pmatrix} p \\ q \\ r \end{pmatrix}.$

In both cases, if det $\mathbf{M} \neq 0$ there is a unique solution to the equations, which can be found by pre-multiplying both sides of the matrix equation by the inverse matrix $\mathbf{M}^{-1}$.

If det $\mathbf{M} = 0$ there is no unique solution to the equations; there may be no solutions or there may be an infinity of solutions.

FUTURE USES

- The work on intersecting planes is developed further in Chapter 8 'Vectors and 3D space'.

8 Vectors and 3D space

> Why is there space rather than no space? Why is space three-dimensional? Why is space big? We have a lot of room to move around in. How come it's not tiny? We have no consensus about these things. We're still exploring them.
>
> Leonard Susskind

1 Finding the angle between two vectors

In this section you will learn how to find the angle between two vectors in two dimensions or three dimensions.

Discussion point
→ Are there any right angles in the building shown above?

Prior knowledge

From Edexcel A level Mathematics Year 1 (AS), you need to be able to use the language of vectors, including the terms magnitude, direction and position vector. You should also be able to find the distance between two points represented by position vectors and be able to add and subtract vectors and multiply a vector by a scalar.

Example 8.1

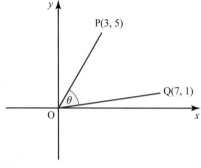

Figure 8.1

Find the angle POQ.

> Remember that $\overrightarrow{OP}$ denotes the vector from O to P, and $\left|\overrightarrow{OP}\right|$ is the *magnitude* (length) of $\overrightarrow{OP}$.

Solution

Using the cosine rule: $\cos\theta = \dfrac{\left|\overrightarrow{OP}\right|^2 + \left|\overrightarrow{OQ}\right|^2 - \left|\overrightarrow{PQ}\right|^2}{2 \times \left|\overrightarrow{OP}\right| \times \left|\overrightarrow{OQ}\right|}$

$\overrightarrow{OP} = \begin{pmatrix} 3 \\ 5 \end{pmatrix}$ so $\left|\overrightarrow{OP}\right| = \sqrt{3^2 + 5^2} = \sqrt{34}$

> Using Pythagoras' theorem.

$\overrightarrow{OQ} = \begin{pmatrix} 7 \\ 1 \end{pmatrix}$ so $\left|\overrightarrow{OQ}\right| = \sqrt{7^2 + 1^2} = \sqrt{50}$

$\overrightarrow{PQ} = \overrightarrow{OQ} - \overrightarrow{OP} = \begin{pmatrix} 7 \\ 1 \end{pmatrix} - \begin{pmatrix} 3 \\ 5 \end{pmatrix} = \begin{pmatrix} 4 \\ -4 \end{pmatrix}$ so $\left|\overrightarrow{PQ}\right| = \sqrt{4^2 + 4^2} = \sqrt{32}$

so $\cos\theta = \dfrac{34 + 50 - 32}{2 \times \sqrt{34} \times \sqrt{50}}$

$\theta = 50.9°$

More generally, to find the angle between $\overrightarrow{OA} = \mathbf{a} = \begin{pmatrix} a_1 \\ a_2 \end{pmatrix}$ and $\overrightarrow{OB} = \mathbf{b} = \begin{pmatrix} b_1 \\ b_2 \end{pmatrix}$ start by applying the cosine rule to the triangle OAB in Figure 8.2.

Discussion point

➔ How else could you find the angle θ?

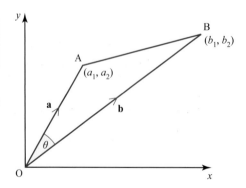

Figure 8.2

$$\cos\theta = \frac{\left|\overrightarrow{OA}\right|^2 + \left|\overrightarrow{OB}\right|^2 - \left|\overrightarrow{AB}\right|^2}{2 \times \left|\overrightarrow{OA}\right| \times \left|\overrightarrow{OB}\right|} \quad \text{①}$$

$\left|\overrightarrow{OA}\right|, \left|\overrightarrow{OB}\right|$ and $\left|\overrightarrow{AB}\right|$ are the lengths of the vectors $\overrightarrow{OA}$, $\overrightarrow{OB}$ and $\overrightarrow{AB}$.

Also from the diagram:

$$\left|\overrightarrow{OA}\right| = |\mathbf{a}| = \sqrt{a_1^2 + a_2^2} \text{ and } \left|\overrightarrow{OB}\right| = |\mathbf{b}| = \sqrt{b_1^2 + b_2^2} \quad \text{②}$$

and

$$\overrightarrow{AB} = \mathbf{b} - \mathbf{a} = \begin{pmatrix} b_1 \\ b_2 \end{pmatrix} - \begin{pmatrix} a_1 \\ a_2 \end{pmatrix} = \begin{pmatrix} b_1 - a_1 \\ b_2 - a_2 \end{pmatrix}$$

so $\left|\overrightarrow{AB}\right| = \sqrt{\left(b_1 - a_1\right)^2 + \left(b_2 - a_2\right)^2}$ ③

ACTIVITY 8.1

By substituting ② and ③ into ① show that $\cos\theta = \dfrac{a_1 b_1 + a_2 b_2}{|\mathbf{a}||\mathbf{b}|}$

where $|\mathbf{a}| = \sqrt{a_1^2 + a_2^2}$ and $|\mathbf{b}| = \sqrt{b_1^2 + b_2^2}$.

The activity above showed that for Figure 8.2, $\cos\theta = \dfrac{a_1 b_1 + a_2 b_2}{|\mathbf{a}||\mathbf{b}|}$.

The expression on the numerator, $a_1 b_1 + a_2 b_2$, is called the scalar product of the vectors $\mathbf{a}$ and $\mathbf{b}$, which is written $\mathbf{a.b}$.

So $\cos\theta = \dfrac{\mathbf{a.b}}{|\mathbf{a}||\mathbf{b}|}$.

This is sometimes called the dot product.

This result is sometimes written $\mathbf{a.b} = |\mathbf{a}||\mathbf{b}|\cos\theta$.

Using the column format, the scalar product can be written as

$$\mathbf{a.b} = \begin{pmatrix} a_1 \\ a_2 \end{pmatrix} . \begin{pmatrix} b_1 \\ b_2 \end{pmatrix} = a_1 b_1 + a_2 b_2.$$

The position vectors of three points A, B and C are given by

Note

1 The scalar product, unlike a vector, has size but no direction.

2 The scalar product of two vectors is *commutative*. This is because multiplication of numbers is commutative. For example:

$$\begin{pmatrix} 3 \\ -4 \end{pmatrix} . \begin{pmatrix} 1 \\ 5 \end{pmatrix} = (3 \times 1) + (-4 \times 5) = (1 \times 3) + (5 \times -4) = \begin{pmatrix} 1 \\ 5 \end{pmatrix} . \begin{pmatrix} 3 \\ -4 \end{pmatrix}$$

The scalar product is found in a similar way for vectors in three dimensions:

$$\begin{pmatrix} a_1 \\ a_2 \\ a_3 \end{pmatrix} \cdot \begin{pmatrix} b_1 \\ b_2 \\ b_3 \end{pmatrix} = a_1 b_1 + a_2 b_2 + a_3 b_3$$

This is used in Example 8.2 to find the angle between two vectors in three dimensions.

Example 8.2

$$\mathbf{a} = \begin{pmatrix} 2 \\ 5 \\ -1 \end{pmatrix}, \mathbf{b} = \begin{pmatrix} 0 \\ 7 \\ 3 \end{pmatrix} \text{ and } \mathbf{c} = \begin{pmatrix} 8 \\ 0 \\ 3 \end{pmatrix}. \text{ Find the vectors } \overrightarrow{AB} \text{ and } \overrightarrow{CB}$$

and hence calculate the angle ABC.

Solution

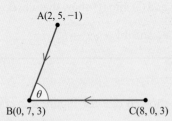

A(2, 5, −1)

B(0, 7, 3)   C(8, 0, 3)

Figure 8.3

$$\overrightarrow{AB} = \mathbf{b} - \mathbf{a} = \begin{pmatrix} 0 \\ 7 \\ 3 \end{pmatrix} - \begin{pmatrix} 2 \\ 5 \\ -1 \end{pmatrix} = \begin{pmatrix} -2 \\ 2 \\ 4 \end{pmatrix}$$

$$\overrightarrow{CB} = \mathbf{b} - \mathbf{c} = \begin{pmatrix} 0 \\ 7 \\ 3 \end{pmatrix} - \begin{pmatrix} 8 \\ 0 \\ 3 \end{pmatrix} = \begin{pmatrix} -8 \\ 7 \\ 0 \end{pmatrix}$$

The angle ABC is found using the scalar product of the vectors $\overrightarrow{AB}$ and $\overrightarrow{CB}$.

$$\overrightarrow{AB} \cdot \overrightarrow{CB} = \begin{pmatrix} -2 \\ 2 \\ 4 \end{pmatrix} \cdot \begin{pmatrix} -8 \\ 7 \\ 0 \end{pmatrix} = 16 + 14 + 0 = 30$$

$$\left| \overrightarrow{AB} \right| = \sqrt{(-2)^2 + 2^2 + 4^2} = \sqrt{24} \text{ and } \left| \overrightarrow{CB} \right| = \sqrt{(-8)^2 + 7^2 + 0^2} = \sqrt{113}$$

$$\overrightarrow{AB} \cdot \overrightarrow{CB} = \left| \overrightarrow{AB} \right| \left| \overrightarrow{CB} \right| \cos\theta$$

$$\Rightarrow 30 = \sqrt{24}\sqrt{113} \cos\theta$$

$$\Rightarrow \cos\theta = \frac{30}{\sqrt{24}\sqrt{113}}$$

$$\Rightarrow \theta = 54.8°$$

Discussion point

➜ For the points A, B and C in Example 8.2, find the scalar product of the vectors $\overrightarrow{BA}$ and $\overrightarrow{BC}$, and comment on your answer.

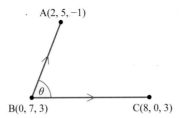

A(2, 5, −1)

B(0, 7, 3)   C(8, 0, 3)

θ

Figure 8.4

Notice that $\overrightarrow{AB}$ and $\overrightarrow{CB}$ are both directed towards the point B, and $\overrightarrow{BA}$ and $\overrightarrow{BC}$ are both directed away from the point B (as in Figure 8.4). Using either pair of vectors gives the angle ABC. This angle could be acute or obtuse.

However, if you use vectors $\overrightarrow{AB}$ (directed towards B) and $\overrightarrow{BC}$ (directed away from B), then you will obtain the angle $180° - \theta$ instead (as in Figure 8.5).

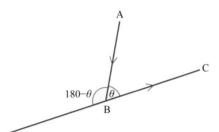

A

C

$180-\theta$ θ

B

Figure 8.5

Perpendicular vectors

If two vectors are perpendicular, then the angle between them is 90°.

Since $\cos 90° = 0$, it follows that if vectors **a** and **b** are perpendicular then $\mathbf{a.b} = 0$.

Conversely, if the scalar product of two non-zero vectors is zero, they are perpendicular.

Example 8.3

Two points, P and Q, have coordinates $(1, 3, -2)$ and $(4, 2, 5)$.
Show that angle POQ = 90°

(i)   using column vectors

(ii) using $\mathbf{i, j, k}$ notation.

Solution

(i) $\mathbf{p} = \begin{pmatrix} 1 \\ 3 \\ -2 \end{pmatrix}, \mathbf{q} = \begin{pmatrix} 4 \\ 2 \\ 5 \end{pmatrix}$

$$\mathbf{p.q} = \begin{pmatrix} 1 \\ 3 \\ -2 \end{pmatrix} \cdot \begin{pmatrix} 4 \\ 2 \\ 5 \end{pmatrix}$$

$$= (1 \times 4) + (3 \times 2) + (-2 \times 5)$$

$$= 4 + 6 - 10$$

$$= 0$$

So the angle POQ = 90°.

(ii) $\mathbf{p} = \mathbf{i} + 3\mathbf{j} - 2\mathbf{k}$, $\mathbf{q} = 4\mathbf{i} + 2\mathbf{j} + 5\mathbf{k}$ | Multiply out the brackets. |

$$\mathbf{p.q} = (\mathbf{i} + 3\mathbf{j} - 2\mathbf{k}).(4\mathbf{i} + 2\mathbf{j} + 5\mathbf{k})$$
$$= 4\mathbf{i.i} + 14\mathbf{i.j} - 3\mathbf{i.k} + 6\mathbf{j.j} + 11\mathbf{j.k} - 10\mathbf{k.k}$$

Since $\mathbf{i}, \mathbf{j}$ and $\mathbf{k}$ are unit vectors, $\mathbf{i.i} = \mathbf{j.j} = \mathbf{k.k} = 1$.

$$= 4 + 6 - 10$$
$$= 0$$

So the angle POQ = 90°.

Since $\mathbf{i}, \mathbf{j}$ and $\mathbf{k}$ are all perpendicular, $\mathbf{i.j} = \mathbf{i.k} = \mathbf{j.k} = 0$.

Exercise 8.1

① Find:

(i) $\begin{pmatrix} 2 \\ 3 \end{pmatrix} \cdot \begin{pmatrix} 1 \\ -2 \end{pmatrix}$ (ii) $\begin{pmatrix} 2 \\ 3 \end{pmatrix} \cdot \begin{pmatrix} -1 \\ 2 \end{pmatrix}$

(iii) $\begin{pmatrix} 1 \\ 2 \\ 3 \end{pmatrix} \cdot \begin{pmatrix} 4 \\ 0 \\ -1 \end{pmatrix}$ (iv) $\begin{pmatrix} 1 \\ 2 \\ 3 \end{pmatrix} \cdot \begin{pmatrix} -1 \\ 4 \\ 0 \end{pmatrix}$

② Find the angle between the vectors $\mathbf{p}$ and $\mathbf{q}$ shown in Figure 8.6.

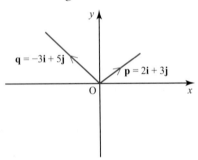

Figure 8.6

③ Find the angle between the vectors:

(i) $\mathbf{a} = 3\mathbf{i} + 2\mathbf{j} - 4\mathbf{k}$ and $\mathbf{b} = -2\mathbf{i} + \mathbf{j} - 3\mathbf{k}$

(ii) $\mathbf{a} = -3\mathbf{i} - 2\mathbf{j} + 4\mathbf{k}$ and $\mathbf{b} = -2\mathbf{i} + \mathbf{j} - 3\mathbf{k}$

(iii) $\mathbf{a} = 3\mathbf{i} + 2\mathbf{j} - 4\mathbf{k}$ and $\mathbf{b} = 2\mathbf{i} - \mathbf{j} + 3\mathbf{k}$

④ Find the angle between the following pairs of vectors and comment on your answers.

(i) $\begin{pmatrix} 3 \\ -2 \\ 5 \end{pmatrix}$ and $\begin{pmatrix} 6 \\ -4 \\ 10 \end{pmatrix}$ (ii) $\begin{pmatrix} 3 \\ -2 \\ 5 \end{pmatrix}$ and $\begin{pmatrix} -9 \\ 6 \\ -15 \end{pmatrix}$

⑤ Find the value of α for which the vectors $\begin{pmatrix} 2 \\ 5 \\ -1 \end{pmatrix}$ and $\begin{pmatrix} 4 \\ -5 \\ \alpha \end{pmatrix}$ are perpendicular.

⑥ Given the vectors $\mathbf{c} = \begin{pmatrix} \alpha \\ 5 \\ 3 \end{pmatrix}$ and $\mathbf{d} = \begin{pmatrix} \alpha \\ \alpha \\ 2 \end{pmatrix}$ are perpendicular, find the possible values of α.

⑦ A triangle has vertices at the points A(2, 1, −3), B(4, 0, 6) and C(−1, 2, 1). Using the scalar product, find the three angles of the triangle ABC and check that they add up to 180°.

⑧ The point A has position vector $\mathbf{a} = \begin{pmatrix} 5 \\ 2 \\ 3 \end{pmatrix}$.

Find the angle that the vector $\mathbf{a}$ makes with each of the coordinate axes.

⑨ The room illustrated in Figure 8.7 has rectangular walls, floor and ceiling. A string has been stretched in a straight line between the corners A and G.

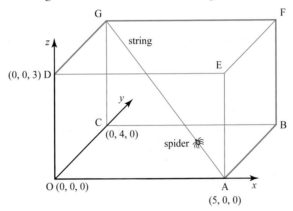

Figure 8.7

The corner O is taken as the origin. A is (5, 0, 0), C is (0, 4, 0) and D is (0, 0, 3), where the lengths are in metres.

A spider walks up the string, starting from A.

(i) Write down the coordinates of G.

(ii) Find the vector $\overrightarrow{AG}$ and the distance the spider walks along the string from A to G.

(iii) Find the angle of elevation of the spider's journey along the string.

⑩ Figure 8.8 shows the design for a barn. Its base and walls are rectangular.

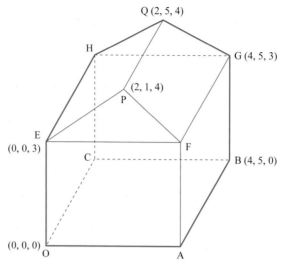

Figure 8.8

(i) Write down the coordinates of the other vertices of the barn.

(ii) Determine whether the section EPF is vertical and hence state the type of quadrilateral formed by the roof sections PFGQ and PQHE.

(iii) Find the cosine of angle FPE and hence find the exact area of the triangle FPE.

The engineer plans to increase the strength of the barn by installing supporting metal bars along OG and AH.

(iv) Calculate the acute angle between the metal bars.

⑪ If $(\mathbf{a} + 2\mathbf{b}) \cdot \mathbf{c} - (3\mathbf{a} + \mathbf{c}) \cdot \mathbf{b} = 5\mathbf{a} \cdot \mathbf{b} - 3\mathbf{a} \cdot \mathbf{c}$ show that $\mathbf{b} \cdot \mathbf{c} = 4\mathbf{a} \cdot (2\mathbf{b} - \mathbf{c})$.

⑫ Three vectors $\mathbf{a}$, $\mathbf{b}$ and $\mathbf{c}$ have magnitudes 5, 2 and 3 respectively.

Using this information, and the properties of the scalar product, simplify $(\mathbf{a} + \mathbf{b} + \mathbf{c}) \cdot \mathbf{a} - (\mathbf{b} + \mathbf{c}) \cdot \mathbf{c} - \mathbf{a} \cdot \mathbf{b}$.

⑬ Two vectors are given by $\mathbf{a} = a_1\mathbf{i} + b_1\mathbf{j} + c_1\mathbf{k}$ and $\mathbf{b} = a_2\mathbf{i} + b_2\mathbf{j} + c_2\mathbf{k}$.

Using the fact that $\mathbf{i} \cdot \mathbf{j} = \mathbf{j} \cdot \mathbf{k} = \mathbf{k} \cdot \mathbf{i} = 0$, show algebraically that $\mathbf{a} \cdot \mathbf{b} = a_1b_1 + a_2b_2 + a_3b_3$.

2 The vector equation of a line

Lines in two dimensions

Before looking at the equation of a line in three dimensions, Activity 8.2 looks at a new format for the equation of a line in two dimensions. This is called the vector equation of the line.

ACTIVITY 8.2

The position vector of a set of points is given by

$$\mathbf{r} = \begin{pmatrix} 2 \\ -1 \end{pmatrix} + \lambda \begin{pmatrix} 2 \\ 4 \end{pmatrix}$$

where λ is a parameter that can take any value and A is the point $(2, -1)$.

(i) Show that $\lambda = 1$ corresponds to the point B with position vector $\begin{pmatrix} 4 \\ 3 \end{pmatrix}$.

(ii) Find the position vectors of the points corresponding to values of λ of $-2, -1,$ $0, \dfrac{1}{2}, \dfrac{3}{4}, 2, 3$.

(iii) Plot the points from parts (i) and (ii) on a sheet of graph paper and show they can be joined to form a straight line.

(iv) What can you say about the position of the point if:

    (a) $0 < \lambda < 1$ (b) $\lambda > 1$ (c) $\lambda < 0$?

This activity should have convinced you that $\mathbf{r} = \begin{pmatrix} 2 \\ -1 \end{pmatrix} + \lambda \begin{pmatrix} 2 \\ 4 \end{pmatrix}$ is the equation of a straight line passing through the point $(2, -1)$. The vector $\begin{pmatrix} 2 \\ 4 \end{pmatrix}$ determines the direction of the line. You might find it helpful to think of this as shown in Figure 8.9.

Starting from the origin, you can get onto the line at a given point A. From here, all other points on the line can then be reached by taking a number of steps (λ steps) in the direction of a given vector, called the **direction vector**.

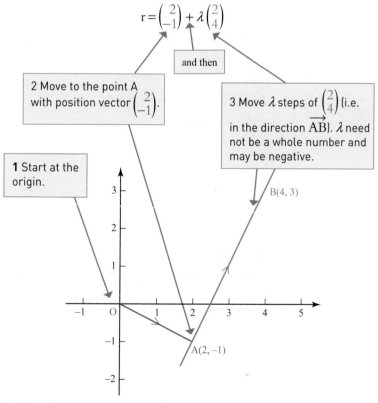

$$r = \begin{pmatrix} 2 \\ -1 \end{pmatrix} + \lambda \begin{pmatrix} 2 \\ 4 \end{pmatrix}$$

and then

2 Move to the point A with position vector $\begin{pmatrix} 2 \\ -1 \end{pmatrix}$.

3 Move λ steps of $\begin{pmatrix} 2 \\ 4 \end{pmatrix}$ (i.e. in the direction $\overrightarrow{AB}$). λ need not be a whole number and may be negative.

1 Start at the origin.

B(4, 3)

A(2, −1)

Figure 8.9

You should also have noticed that:

$\lambda = 0$ corresponds to the point A

$\lambda = 1$ corresponds to the point B, one 'step' along the line away from A

$0 < \lambda < 1$ corresponds to points lying between A and B

$\lambda > 1$ corresponds to points lying beyond B

$\lambda < 0$ corresponds to points beyond A, in the opposite direction to B.

The vector equation of a line is not unique. In this case, any vector parallel, or in the opposite direction to, $\begin{pmatrix} 2 \\ 4 \end{pmatrix}$ could be used as the direction vector, for example, $\begin{pmatrix} 1 \\ 2 \end{pmatrix}, \begin{pmatrix} 3 \\ -6 \end{pmatrix}$ or $\begin{pmatrix} 20 \\ 40 \end{pmatrix}$. Similarly, you can 'step' on to the line at any point, such as B(4, 3).

> In two dimensions the equation of a line usually looks easier in Cartesian form than in vector form. However, as you are about to see, the opposite is the case in three dimensions where the vector form is much easier to work with.

So the line $\mathbf{r} = \begin{pmatrix} 2 \\ -1 \end{pmatrix} + \lambda \begin{pmatrix} 2 \\ 4 \end{pmatrix}$ could also, for example, have equation

$\mathbf{r} = \begin{pmatrix} 2 \\ -1 \end{pmatrix} + \lambda \begin{pmatrix} 1 \\ 2 \end{pmatrix}$ or $\mathbf{r} = \begin{pmatrix} 4 \\ 3 \end{pmatrix} + \lambda \begin{pmatrix} 2 \\ 4 \end{pmatrix}$.

In general, the **vector equation of a line** in two dimensions is given by:

$$\mathbf{r} = \mathbf{a} + \lambda \mathbf{d}$$

where $\mathbf{a}$ is the position vector of a point A on the line and $\mathbf{d}$ is the direction vector of the line. Sometimes a different letter, such as μ or t is used as the parameter instead of λ.

ACTIVITY 8.3

The vector equation of a line

$$\mathbf{r} = \mathbf{a} + \lambda \mathbf{d}$$

is written in the form

$$\begin{pmatrix} x \\ y \end{pmatrix} = \begin{pmatrix} a_1 \\ a_2 \end{pmatrix} + \lambda \begin{pmatrix} d_1 \\ d_2 \end{pmatrix}$$

(i) Write down expressions for x and y in terms of λ.

(ii) Rearrange the two expressions from part (i) to make λ the subject. By equating these two expressions, show that the vector equation of the line can be written in the form

$$y = mx + c$$

where m and c are constants.

Activity 8.3 shows that the vector and Cartesian equations of a line are equivalent.

Lines in three dimensions

The same form for the vector equation of a line can be used in three dimensions. For example:

$$\mathbf{r} = \begin{pmatrix} 3 \\ 4 \\ 1 \end{pmatrix} + \lambda \begin{pmatrix} 2 \\ 3 \\ 6 \end{pmatrix}$$

represents a line through the point with position vector $\begin{pmatrix} 3 \\ 4 \\ 1 \end{pmatrix}$ with direction vector $\begin{pmatrix} 2 \\ 3 \\ 6 \end{pmatrix}$.

Writing $\mathbf{r}$ as $\begin{pmatrix} x \\ y \\ z \end{pmatrix}$ gives $\begin{pmatrix} x \\ y \\ z \end{pmatrix} = \begin{pmatrix} 3 \\ 4 \\ 1 \end{pmatrix} + \lambda \begin{pmatrix} 2 \\ 3 \\ 6 \end{pmatrix}$.

This equation contains the three relationships

$$x = 3 + 2\lambda \qquad y = 4 + 3\lambda \qquad z = 1 + 6\lambda$$

Making λ the subject of each of these gives:

> This form is not easy to work with and you will often find that the first step in a problem is to convert the Cartesian form into vector form.

$$\lambda = \frac{x - 3}{2} = \frac{y - 4}{3} = \frac{z - 1}{6}$$

This is the **Cartesian equation of a line** in three dimensions.

Generally, a line with direction vector $\mathbf{d} = \begin{pmatrix} d_1 \\ d_2 \\ d_3 \end{pmatrix}$ passing through the point A with position vector $\mathbf{a} = \begin{pmatrix} a_1 \\ a_2 \\ a_3 \end{pmatrix}$ has the Cartesian equation

> The direction vector of the line can be read from the denominators of the three expressions in this equation; the point A can be determined from the three numerators.

$$\frac{x - a_1}{d_1} = \frac{y - a_2}{d_2} = \frac{z - a_3}{d_3}$$

Special cases of the Cartesian equation of a line

In the equation

$$\frac{x - a_1}{d_1} = \frac{y - a_2}{d_2} = \frac{z - a_3}{d_3}$$

it is possible that one or two of the values of d_i might equal zero. In such cases the equation of the line needs to be written differently.

For example, the line through $(7, 2, 3)$ in the direction $\begin{pmatrix} 0 \\ 5 \\ 2 \end{pmatrix}$ would have equation

$$\lambda = \frac{x - 7}{0} = \frac{y - 2}{5} = \frac{z - 3}{2}$$

The first fraction involves division by zero, which is undefined. The expression $\frac{x - 7}{0}$ comes from the rearrangement of $x - 7 = 0\lambda$ and so you can write this as $x - 7 = 0$ or $x = 7$.

So the equation of this line would be written

$$x = 7 \text{ and } \lambda = \frac{y - 2}{5} = \frac{z - 3}{2}$$

Note that the three d_i cannot all equal zero as it is not possible for a line to have a zero direction vector.

Example 8.4

Write the equation of this line in vector form:

$$\frac{x-4}{3} = \frac{y-3}{-2} = \frac{z+6}{4}$$

Solution

$$\lambda = \frac{x-4}{3} = \frac{y-3}{-2} = \frac{z+6}{4}$$

$$\lambda = \frac{x-4}{3} \Rightarrow x = 3\lambda + 4$$

$$\lambda = \frac{y-3}{-2} \Rightarrow y = -2\lambda + 3$$

$$\lambda = \frac{z+6}{4} \Rightarrow z = 4\lambda - 6$$

So $\mathbf{r} = \begin{pmatrix} x \\ y \\ z \end{pmatrix} = \begin{pmatrix} 3\lambda + 4 \\ -2\lambda + 3 \\ 4\lambda - 6 \end{pmatrix} \Rightarrow \mathbf{r} = \begin{pmatrix} 4 \\ 3 \\ -6 \end{pmatrix} + \lambda \begin{pmatrix} 3 \\ -2 \\ 4 \end{pmatrix}$

Example 8.5

Find the Cartesian form of the equation of the line through the point $A(7, -12, 4)$ in the direction $2\mathbf{i} - 5\mathbf{j} - 3\mathbf{k}$.

Solution

The line has vector form $\mathbf{r} = \mathbf{a} + \lambda \mathbf{d} = \begin{pmatrix} 7 \\ -12 \\ 4 \end{pmatrix} + \lambda \begin{pmatrix} 2 \\ -5 \\ -3 \end{pmatrix}$.

This leads to the equations

$$x = 7 + 2\lambda \qquad y = -12 - 5\lambda \qquad z = 4 - 3\lambda$$

which can be rearranged to give the Cartesian equation

$$\lambda = \frac{x-7}{2} = \frac{y+12}{-5} = \frac{z-4}{-3}$$

The intersection of straight lines in two dimensions

You already know how to find the point of intersection of two straight lines given in Cartesian form, by using simultaneous equations.

You can also use vector methods to find the position vector of the point where two lines intersect.

Example 8.6

Notice that different letters are used for the parameters in the two equations to avoid confusion.

Find the position vector of the point where the following lines intersect.

$$\mathbf{r} = \begin{pmatrix} 2 \\ 3 \end{pmatrix} + \lambda \begin{pmatrix} 1 \\ 2 \end{pmatrix} \text{ and } \mathbf{r} = \begin{pmatrix} 6 \\ 1 \end{pmatrix} + \mu \begin{pmatrix} 1 \\ -3 \end{pmatrix}$$

Solution

When the lines intersect, the position vector is the same for each of them.

$$\mathbf{r} = \begin{pmatrix} x \\ y \end{pmatrix} = \begin{pmatrix} 2 \\ 3 \end{pmatrix} + \lambda \begin{pmatrix} 1 \\ 2 \end{pmatrix} = \begin{pmatrix} 6 \\ 1 \end{pmatrix} + \mu \begin{pmatrix} 1 \\ -3 \end{pmatrix}$$

This gives two simultaneous equations for λ and μ.

$$2 + \lambda = 6 + \mu$$

$$3 + 2\lambda = 1 - 3\mu$$

Solving these gives $\lambda = 2$ and $\mu = -2$. Substituting in either equation gives

$$\mathbf{r} = \begin{pmatrix} 4 \\ 7 \end{pmatrix}$$

which is the position vector of the point of intersection.

The intersection of straight lines in three dimensions

Hold a pen and a pencil to represent two distinct straight lines as follows:

- hold them to represent two parallel lines;
- hold them to represent two lines intersecting at a unique point;
- hold them to represent lines which are not parallel and which do not intersect even if you were to extend them.

In three-dimensional space, two or more straight lines which are not parallel and which do not meet are known as **skew** lines. In two dimensions, two distinct lines are either parallel or intersecting but in three dimensions there are three possibilities: the lines are either parallel, intersecting or skew. This is illustrated in the following examples.

Example 8.7

The lines l_1 and l_2 are represented by the equations

$$l_1 : \frac{x-1}{1} = \frac{y+6}{2} = \frac{z+1}{3} \qquad l_2 : \frac{x-9}{2} = \frac{y-7}{3} = \frac{z-2}{-1}.$$

(i)   Write these lines in vector form.

(ii) Hence find whether the lines meet, and, if so, the coordinates of their point of intersection.

Solution

(i)   The equation of l_1 is $\mathbf{r} = \begin{pmatrix} 1 \\ -6 \\ -1 \end{pmatrix} + \lambda \begin{pmatrix} 1 \\ 2 \\ 3 \end{pmatrix}$

The equation of l_2 is $\mathbf{r} = \begin{pmatrix} 9 \\ 7 \\ 2 \end{pmatrix} + \mu \begin{pmatrix} 2 \\ 3 \\ -1 \end{pmatrix}$

(ii) If there is a point $\begin{pmatrix} X \\ Y \\ Z \end{pmatrix}$ that is common to both lines then

$$\begin{pmatrix} X \\ Y \\ Z \end{pmatrix} = \begin{pmatrix} 1 \\ -6 \\ -1 \end{pmatrix} + \lambda \begin{pmatrix} 1 \\ 2 \\ 3 \end{pmatrix} = \begin{pmatrix} 9 \\ 7 \\ 2 \end{pmatrix} + \mu \begin{pmatrix} 2 \\ 3 \\ -1 \end{pmatrix}$$

for some parameters λ and μ.

This gives the three equations

$$X = \lambda + 1 = 2\mu + 9 \qquad ①$$
$$Y = 2\lambda - 6 = 3\mu + 7 \qquad ②$$
$$Z = 3\lambda - 1 = -\mu + 2 \qquad ③$$

Now solve any two of the three equations simultaneously.

$$\left. \begin{array}{r} \lambda - 2\mu = 8 \\ 2\lambda - 3\mu = 13 \end{array} \right\} \Leftrightarrow \left. \begin{array}{r} 2\lambda - 4\mu = 16 \\ 2\lambda - 3\mu = 13 \end{array} \right\} \Leftrightarrow \mu = -3, \lambda = 2$$

> Using ① and ②.

If these values for λ and μ also satisfy equation ③, then the lines meet.

Using equation ③, when $\lambda = 2$, $Z = 6 - 1 = 5$ and when $\mu = -3$, $Z = 3 + 2 = 5$.

As both values of Z are equal this proves the lines intersect.

Using either $\lambda = 2$ or $\mu = -3$ in equations ①, ② and ③ gives $X = 3$, $Y = -2$, $Z = 5$ so the lines meet at the point $(3, -2, 5)$.

Example 8.8

Prove that the lines l_1 and l_2 are skew, where:

$$l_1 : \mathbf{r} = \begin{pmatrix} 1 \\ -6 \\ -1 \end{pmatrix} + \lambda \begin{pmatrix} 1 \\ 2 \\ 3 \end{pmatrix}$$

$$l_2 : \mathbf{r} = \begin{pmatrix} 9 \\ 8 \\ 2 \end{pmatrix} + \mu \begin{pmatrix} 2 \\ 3 \\ -1 \end{pmatrix}$$

Solution

If there is a point (X, Y, Z) common to both lines then

$$\begin{pmatrix} X \\ Y \\ Z \end{pmatrix} = \begin{pmatrix} 1 \\ -6 \\ -1 \end{pmatrix} + \lambda \begin{pmatrix} 1 \\ 2 \\ 3 \end{pmatrix} = \begin{pmatrix} 9 \\ 8 \\ 2 \end{pmatrix} + \mu \begin{pmatrix} 2 \\ 3 \\ -1 \end{pmatrix}$$

for some parameters λ and μ.

$X = \lambda + 1 = 2\mu + 9$ ①

$Y = 2\lambda - 6 = 3\mu + 8$ ②

$Z = 3\lambda - 1 = -\mu + 2$ ③

> Solving equations ① and ② simultaneously.

$\left.\begin{array}{l} \lambda - 2\mu = 8 \\ 2\lambda - 3\mu = 14 \end{array}\right\} \Leftrightarrow \left.\begin{array}{l} 2\lambda - 4\mu = 16 \\ 2\lambda - 3\mu = 14 \end{array}\right\} \Leftrightarrow \mu = -2, \lambda = 4$

When $\lambda = 4$, $Z = 12 - 1 = 11$

and when $\mu = -2$,

$Z = 2 + 2 = 4$.

> Substitute these values into equation ③.

Therefore, the values $\mu = -2$, $\lambda = 4$ do not satisfy the third equation and so the lines do not meet. As the lines are distinct, the only other alternatives are that the lines are parallel or skew.

Look at the direction vectors of the lines: $\begin{pmatrix} 1 \\ 2 \\ 3 \end{pmatrix}$ and $\begin{pmatrix} 2 \\ 3 \\ -1 \end{pmatrix}$. Neither of

these is a multiple of the other so they are not parallel and hence the two lines are not parallel. So, lines l_1 and l_2 are skew.

Finding the angle between two lines

Figure 8.10 shows two lines in two dimensions, with their equations given in vector form.

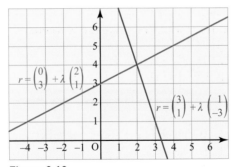

Figure 8.10

The angle between the two lines is the same as the angle between their direction

vectors, $\begin{pmatrix} 2 \\ 1 \end{pmatrix}$ and $\begin{pmatrix} 1 \\ -3 \end{pmatrix}$. So you can use the scalar product to find the angle

between the two lines.

Example 8.9

Find the acute angle between the lines $\mathbf{r} = \begin{pmatrix} 0 \\ 3 \end{pmatrix} + \lambda \begin{pmatrix} 2 \\ 1 \end{pmatrix}$

and $\mathbf{r} = \begin{pmatrix} 3 \\ 1 \end{pmatrix} + \mu \begin{pmatrix} 1 \\ -3 \end{pmatrix}$.

ACTIVITY 8.4

Find the Cartesian forms of the two equations in Example 8.9. How can you find the angle between them without using vectors?

Line m is in front of line l, so m' has been moved 'into' the page in order to intersect with line l.

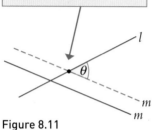

Figure 8.11

Solution

$$\begin{pmatrix} 2 \\ 1 \end{pmatrix} \cdot \begin{pmatrix} 1 \\ -3 \end{pmatrix} = (2 \times 1) + (1 \times -3) = 2 - 3 = -1$$

$$\left| \begin{pmatrix} 2 \\ 1 \end{pmatrix} \right| = \sqrt{2^2 + 1^2} = \sqrt{5} \text{ and } \left| \begin{pmatrix} 1 \\ -3 \end{pmatrix} \right| = \sqrt{1^2 + (-3)^2} = \sqrt{10}$$

$$\cos\theta = \frac{-1}{\sqrt{5}\sqrt{10}}$$

$$\theta = 98.1°$$

So the acute angle between the lines is $180° - 98.1° = 81.9°$

The same method can be used for lines in three dimensions. Even if the lines do not meet, the angle between them is still the angle between their direction vectors.

The lines l and m shown in Figure 8.11 are skew. The angle between them is shown by the angle θ between the lines l and m', where m' is a translation of the line m to a position where it intersects the line l.

Example 8.10

Find the angle between the lines $\mathbf{r} = \begin{pmatrix} 1 \\ 0 \\ 4 \end{pmatrix} + \lambda \begin{pmatrix} 2 \\ -1 \\ -1 \end{pmatrix}$ and

$$\mathbf{r} = \begin{pmatrix} 2 \\ -1 \\ 3 \end{pmatrix} + \mu \begin{pmatrix} 3 \\ 0 \\ 1 \end{pmatrix}.$$

Solution

The angle between the lines is the angle between their direction vectors

$$\begin{pmatrix} 2 \\ -1 \\ -1 \end{pmatrix} \text{ and } \begin{pmatrix} 3 \\ 0 \\ 1 \end{pmatrix}.$$

Using $\mathbf{a} \cdot \mathbf{b} = |\mathbf{a}||\mathbf{b}|\cos\theta$,

$$\begin{pmatrix} 2 \\ -1 \\ -1 \end{pmatrix} \cdot \begin{pmatrix} 3 \\ 0 \\ 1 \end{pmatrix} = \sqrt{2^2 + (-1)^2 + (-1)^2} \sqrt{3^2 + 0^2 + 1^2} \cos\theta$$

$$\Rightarrow 6 + 0 - 1 = \sqrt{6}\sqrt{10}\cos\theta$$

$$\Rightarrow \theta = \cos^{-1}\left(\frac{5}{\sqrt{6}\sqrt{10}}\right) = 49.8°$$

① Find the equation of the following lines in vector form:

    (i)   through $(3, 1)$ in the direction $\begin{pmatrix} 5 \\ -2 \end{pmatrix}$

    (ii) through $(5, -1)$ in the direction $\begin{pmatrix} 0 \\ 4 \end{pmatrix}$

    (iii) through $(-2, 4)$ and $(3, 9)$

    (iv) through $(0, 8)$ and $(-2, -3)$.

② Find the equation of the following lines in vector form:

    (i)   through $(2, 4, -1)$ in the direction $\begin{pmatrix} 3 \\ 6 \\ 4 \end{pmatrix}$

    (ii) through $(1, 0, -1)$ in the direction $\begin{pmatrix} 1 \\ 0 \\ 0 \end{pmatrix}$

    (iii) through $(1, 0, 4)$ and $(6, 3, -2)$

    (iv) through $(0, 0, 1)$ and $(2, 1, 4)$.

③ Write the equations of the following lines in Cartesian form.

    (i) $\mathbf{r} = \begin{pmatrix} 2 \\ 4 \\ -1 \end{pmatrix} + t\begin{pmatrix} 3 \\ 6 \\ 4 \end{pmatrix}$ (ii) $\mathbf{r} = \begin{pmatrix} 1 \\ 0 \\ -1 \end{pmatrix} + t\begin{pmatrix} 1 \\ 3 \\ 4 \end{pmatrix}$

    (iii) $\mathbf{r} = \begin{pmatrix} 3 \\ 0 \\ 4 \end{pmatrix} + t\begin{pmatrix} 1 \\ 0 \\ 2 \end{pmatrix}$ (iv) $\mathbf{r} = \begin{pmatrix} 0 \\ 4 \\ 1 \end{pmatrix} + t\begin{pmatrix} 2 \\ 0 \\ 4 \end{pmatrix}$

④ Write the equations of the following lines in vector form.

    (i) $\dfrac{x-3}{5} = \dfrac{y+2}{3} = \dfrac{z-1}{4}$ (ii) $x = \dfrac{y}{2} = \dfrac{z+1}{3}$

    (iii) $x = y = z$ (iv) $x = 2$ and $y = z$

⑤ Write down the vector and Cartesian equations of the line through the point $(3, -5, 2)$ which is parallel to the y-axis.

⑥ Find the position vector of the point of intersection of each of these pairs of lines.

    (i) $\mathbf{r} = \begin{pmatrix} 2 \\ 1 \end{pmatrix} + \lambda\begin{pmatrix} 1 \\ 0 \end{pmatrix}$ and $\mathbf{r} = \begin{pmatrix} 3 \\ 0 \end{pmatrix} + \mu\begin{pmatrix} 1 \\ 1 \end{pmatrix}$

    (ii) $\mathbf{r} = \begin{pmatrix} 2 \\ -1 \end{pmatrix} + \lambda\begin{pmatrix} 1 \\ 2 \end{pmatrix}$ and $\mathbf{r} = \mu\begin{pmatrix} 1 \\ 1 \end{pmatrix}$

    (iii) $\mathbf{r} = \begin{pmatrix} -2 \\ -3 \end{pmatrix} + \lambda\begin{pmatrix} -1 \\ 3 \end{pmatrix}$ and $\mathbf{r} = \begin{pmatrix} 1 \\ 3 \end{pmatrix} + \mu\begin{pmatrix} 2 \\ -1 \end{pmatrix}$

⑦ Decide whether the following pairs of lines intersect or not. If they do intersect, find the point of intersection; if not, state whether the lines are parallel or skew.

(i) $L_1 : \dfrac{x-6}{1} = \dfrac{y+4}{-2} = \dfrac{z-2}{5}$ $\qquad L_2 : \dfrac{x-1}{1} = \dfrac{y-4}{-1} = \dfrac{z+17}{2}$

(ii) $L_1 : \dfrac{x}{5} = \dfrac{y+1}{3} = \dfrac{z-4}{-3}$ $\qquad L_2 : \dfrac{x-2}{4} = \dfrac{y-5}{-3} = \dfrac{z+1}{2}$

(iii) $\mathbf{r}_1 = \begin{pmatrix} 2 \\ 0 \\ 1 \end{pmatrix} + \lambda \begin{pmatrix} 3 \\ 2 \\ 1 \end{pmatrix}$ $\qquad \mathbf{r}_2 = \begin{pmatrix} 4 \\ 9 \\ -1 \end{pmatrix} + \mu \begin{pmatrix} -6 \\ -4 \\ -2 \end{pmatrix}$

(iv) $\mathbf{r}_1 = \begin{pmatrix} 9 \\ 3 \\ -4 \end{pmatrix} + \lambda \begin{pmatrix} 1 \\ 2 \\ -3 \end{pmatrix}$ $\qquad \mathbf{r}_2 = \begin{pmatrix} 1 \\ -4 \\ 5 \end{pmatrix} + \mu \begin{pmatrix} 1 \\ -1 \\ 2 \end{pmatrix}$

(v) $\mathbf{r}_1 = \begin{pmatrix} 2 \\ 3 \\ 1 \end{pmatrix} + \lambda \begin{pmatrix} 1 \\ 1 \\ -2 \end{pmatrix}$ $\qquad \mathbf{r}_2 = \begin{pmatrix} -1 \\ -3 \\ -1 \end{pmatrix} + \mu \begin{pmatrix} 1 \\ 3 \\ 2 \end{pmatrix}$

⑧ Find the acute angle between these pairs of lines

(i) $\mathbf{r} = \begin{pmatrix} 2 \\ 5 \end{pmatrix} + \lambda \begin{pmatrix} 1 \\ 2 \end{pmatrix}$ and $\mathbf{r} = \begin{pmatrix} 1 \\ 2 \end{pmatrix} + \mu \begin{pmatrix} -1 \\ 3 \end{pmatrix}$

(ii) $\mathbf{r} = \begin{pmatrix} 0 \\ 3 \end{pmatrix} + \lambda \begin{pmatrix} -5 \\ 1 \end{pmatrix}$ and $\mathbf{r} = \begin{pmatrix} 2 \\ -1 \end{pmatrix} + \mu \begin{pmatrix} 1 \\ 1 \end{pmatrix}$

(iii) $\mathbf{r} = \begin{pmatrix} 2 \\ 1 \\ 3 \end{pmatrix} + \lambda \begin{pmatrix} 1 \\ 4 \\ 0 \end{pmatrix}$ and $\mathbf{r} = \begin{pmatrix} 6 \\ 10 \\ 4 \end{pmatrix} + \mu \begin{pmatrix} 2 \\ 1 \\ 1 \end{pmatrix}$

(iv) $\mathbf{r} = \lambda \begin{pmatrix} 4 \\ 1 \\ 4 \end{pmatrix}$ and $\mathbf{r} = \begin{pmatrix} 7 \\ 0 \\ -3 \end{pmatrix} + \mu \begin{pmatrix} 1 \\ 2 \\ -1 \end{pmatrix}$

(v) $\dfrac{x-4}{3} = \dfrac{y-2}{7} = \dfrac{z+1}{-4}$ and $\dfrac{x-5}{2} = \dfrac{y-1}{8} = \dfrac{z}{-5}$

⑨ To support a tree damaged in a gale a tree surgeon attaches wire ropes to four of the branches, as shown in Figure 8.12.

Figure 8.12

He joins $(2, 0, 3)$ to $(-1, 2, 6)$ and $(0, 3, 5)$ to $(-2, -2, 4)$.

Do the ropes, assumed to be straight, meet?

⑩ Show that the lines

$$L_1 : \frac{x+7}{4} = \frac{y-24}{-7} = \frac{z+4}{4}$$

$$L_2 : \frac{x-3}{2} = \frac{y+10}{2} = \frac{z-15}{-1}$$

$$L_3 : \frac{x+3}{8} = \frac{y-6}{-3} = \frac{z-6}{2}$$

form a triangle and find the length of its sides.

⑪ Figure 8.13 shows a music stand, consisting of a rectangle DEFG with a vertical support OA.

Relative to axes through the origin O, which is on the floor, the coordinates of various points are given, with dimensions in metres, as A(0, 0, 1), D(−0.25, 0, 1) and F(0.25, 0.15, 1.3).

DE and GF are horizontal. A is the midpoint of DE, and B is the midpoint of GF.

C is on AB so that $AC = \frac{1}{3}AB$.

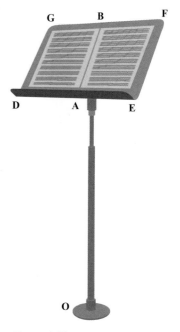

Figure 8.13

(i) Write down the vector $\overrightarrow{AD}$ and show that $\overrightarrow{EF}$ is $\begin{pmatrix} 0 \\ 0.15 \\ 0.3 \end{pmatrix}$.

(ii) Calculate the coordinates of C.

(iii) Find the equations of the lines DE and EF in vector form.

⑫ Figure 8.14 illustrates the flight path of a helicopter H taking off from an airport.

The origin O is situated at the base of the airport control tower, the x-axis is due east, the y-axis due north and the z-axis vertical.

The units of distance are kilometres.

The helicopter takes off from the point G.

The position vector **r** of the helicopter t minutes after take-off is given by

$$\mathbf{r} = (1+t)\mathbf{i} + (0.5+2t)\mathbf{j} + 2t\mathbf{k}$$

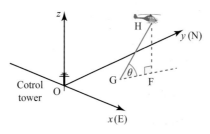

Figure 8.14

(i)   Write down the coordinates of G.

(ii) Find the angle the flight path makes with the horizontal (this is shown as angle θ in the diagram).

(iii) Find the bearing of the flight path (i.e. the bearing of the line GF).

The helicopter enters a cloud at a height of 2 km.

(iv) Find the coordinates of the point where the helicopter enters the cloud.

A mountain top is situated at M(5, 4.5, 3).

(v)   Find the value of t when HM is perpendicular to the flight path GH.

Find the distance from the helicopter to the mountain top at this time.

3 The equation of a plane

You can write the equation of a plane in either vector or Cartesian form. The Cartesian form is used more often but to see where it comes from it is helpful to start with the vector form.

> ## Discussion points
>
> Lay a sheet of paper on a flat horizontal table and mark several straight lines on it. Now take a pencil and stand it upright on the sheet of paper (see Figure 8.15).
>
>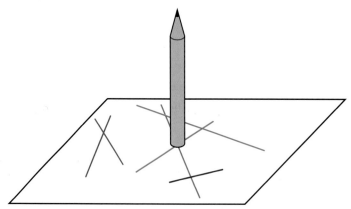
>
> Figure 8.15
>
> ➜ What angle does the pencil make with any individual line?
>
> ➜ Would it make any difference if the table were tilted at an angle (apart from the fact that you could no longer balance the pencil)?

The discussion above shows you that there is a direction (that of the pencil) which is at right angles to every straight line in the plane. A line in that direction is said to be perpendicular to the plane and is referred to as a **normal** to the plane.

It is often denoted by $\mathbf{n} = \begin{pmatrix} n_1 \\ n_2 \\ n_3 \end{pmatrix}$.

In Figure 8.16 the point A is on the plane and the vector **n** is perpendicular to the plane. This information allows you to find an expression for the position vector **r** of a general point R on the plane.

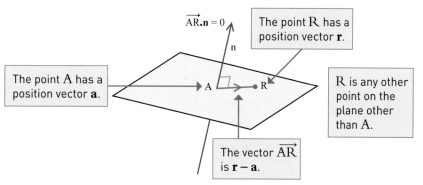

Figure 8.16

The vector $\overrightarrow{AR}$ is a line in the plane, and so it follows that $\overrightarrow{AR}$ is at right angles to the direction **n**.

$$\overrightarrow{AR}\cdot\mathbf{n} = 0$$

The vector $\overrightarrow{AR}$ is given by $\overrightarrow{AR} = \mathbf{r} - \mathbf{a}$ and so

$(\mathbf{r} - \mathbf{a})\cdot\mathbf{n} = 0.$ ◄─ This can also be written as **r·n** = **k** where **k** is the value of **a.n**.

This is the vector equation of the plane.

Expanding the brackets lets you write this in an alternative form as

$$\mathbf{r}\cdot\mathbf{n} - \mathbf{a}\cdot\mathbf{n} = 0.$$

Although the vector equation of a plane is very compact, it is more common to use the Cartesian form. This is derived from the vector form as follows.

Write the normal vector **n** as $\begin{pmatrix} n_1 \\ n_2 \\ n_3 \end{pmatrix}$ and the position vector of A as

$\mathbf{a} = \begin{pmatrix} a_1 \\ a_2 \\ a_3 \end{pmatrix}$. The position vector of the general point R is $\mathbf{r} = \begin{pmatrix} x \\ y \\ z \end{pmatrix}$.

TECHNOLOGY

If you have access to 3D graphing software, experiment with planes in the form $ax + by + cz + d = 0$, varying the values of a, b, c and d.

So the equation $\mathbf{r}\cdot\mathbf{n} - \mathbf{a}\cdot\mathbf{n} = 0$

can be written as $\begin{pmatrix} x \\ y \\ z \end{pmatrix} \cdot \begin{pmatrix} n_1 \\ n_2 \\ n_3 \end{pmatrix} - \begin{pmatrix} a_1 \\ a_2 \\ a_3 \end{pmatrix} \cdot \begin{pmatrix} n_1 \\ n_2 \\ n_3 \end{pmatrix} = 0.$

Notice that d is a constant and is a scalar.

This is the same as $n_1 x + n_2 y + n_3 z + d = 0$ where $d = -(a_1 n_1 + a_2 n_2 + a_3 n_3)$.

The following example shows you how to use this.

Example 8.11

The point A $(2, 3, -5)$ lies on a plane. The vector $\mathbf{n} = \begin{pmatrix} -4 \\ 2 \\ 1 \end{pmatrix}$ is perpendicular to the plane.

(i) Find the Cartesian equation of the plane.

(ii) Investigate whether the points $P(5, 3, -2)$ and $Q(3, 5, -5)$ lie in the plane.

Solution

(i) The Cartesian equation of the plane is
$$n_1 x + n_2 y + n_3 z + d = 0.$$

$$\mathbf{n} = \begin{pmatrix} -4 \\ 2 \\ 1 \end{pmatrix} \text{ so } n_1 = -4, \ n_2 = 2 \text{ and } n_3 = 1$$

The equation of the plane is $-4x + 2y + z + d = 0$.

It remains to find d. There are two ways of doing this.

Either:
The point A is $(2, 3, -5)$.
Substituting for x, y and z in
$-4x + 2y + z + d = 0$ gives

$-4 \times 2 + 2 \times 3 - 5 + d = 0$

so $d = 8 - 6 + 5 = 7$.

Or:
$d = -\mathbf{a}.\mathbf{n}$
where $\mathbf{a}$ is the position vector of A, $(2, 3, -5)$.

So $\mathbf{a} = \begin{pmatrix} 2 \\ 3 \\ -5 \end{pmatrix}$ and $\mathbf{n} = \begin{pmatrix} -4 \\ 2 \\ 1 \end{pmatrix}$

$$d = -\begin{pmatrix} 2 \\ 3 \\ -5 \end{pmatrix} \cdot \begin{pmatrix} -4 \\ 2 \\ 1 \end{pmatrix}$$
$$= -[(2 \times -4) + (3 \times 2) + (-5 \times 1)]$$
$$= -[-8 + 6 - 5] = 7$$

So the equation of the plane is $-4x + 2y + z + 7 = 0$.

(ii) P is $(5, 3, -2)$.
Substituting in the left-hand side of the equation of the plane gives
$(-4 \times 5) + (2 \times 3) - 2 + 7 = -9$.
Since this is not equal to 0, P does not lie in the plane.
Q is $(3, 5, -5)$.
Substituting in the left-hand side of the equation of the plane gives
$(-4 \times 3) + (2 \times 5) - 5 + 7 = 0$.
Since this is equal to 0, Q lies in the plane.

Look carefully at the equation of the plane in this example. You can see at once that the vector $\begin{pmatrix} -4 \\ 2 \\ 1 \end{pmatrix}$, formed from the coefficients of x, y and z, is perpendicular to the plane.

In general the vector $\begin{pmatrix} n_1 \\ n_2 \\ n_3 \end{pmatrix}$ is perpendicular to all planes of the form

$n_1 x + n_2 y + n_3 z + d = 0$, whatever the value of d (see Figure 8.17).

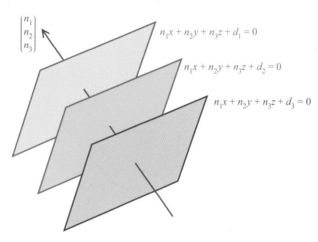

Figure 8.17

Consequently, all planes of that form are parallel; the coefficients of x, y and z determine the direction of the plane, the value of d its location.

Example 8.12

Find the Cartesian equation of the plane which is parallel to the plane $3x - y + 2z + 5 = 0$ and contains the point $(1, 0, -2)$.

Solution

The normal to the plane $3x - y + 2z + 5 = 0$ is $\begin{pmatrix} 3 \\ -1 \\ 2 \end{pmatrix}$.

Any plane parallel to this plane has the same normal vector, so the required plane has equation of the form $3x - y + 2z + d = 0$.
The plane contains the point $(1, 0, -2)$, so

$(3 \times 1) - 0 + (2 \times -2) + d = 0$

$\Rightarrow d = 1$

The equation of the plane is $3x - y + 2z + 1 = 0$.

Discussion point

➔ Given the coordinates of three points A, B and C in a plane, how could you find the equation of the plane?

Notation

So far the Cartesian equation of a plane has been written as
$$n_1 x + n_2 y + n_3 z + d = 0.$$
Another common way of writing it is $ax + by + cz + d = 0$.

In this case the vector $\begin{pmatrix} a \\ b \\ c \end{pmatrix}$ is normal to the plane.

ACTIVITY 8.5

A plane $ax + by + cz + d = 0$ contains the points $(1, 1, 1)$, $(1, -1, 0)$ and $(-1, 0, 2)$.

Use this information to write down three simultaneous equations and use a matrix method to solve these. Hence find the equation of the plane.

Sometimes you may need to find the equation of a plane that contains three given points.

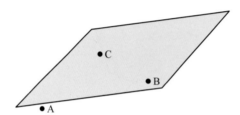

Figure 8.18

For example, suppose that $A = (3, -2, 5)$, $B = (1, -6, -1)$ and $C = (9, 7, 2)$.

You could write the equation of the plane as $n_1 x + n_2 y + n_3 z + d = 0$ and substitute the coordinates of A, B and C to get three simultaneous equations:

$$3n_1 - 2n_2 + 5n_3 + d = 0, \quad n_1 - 6n_2 - n_3 + d = 0, \quad 9n_1 + 7n_2 + 2n_3 + d = 0$$

However there are more unknowns than equations so the equations cannot be solved to find unique values for n_1, n_2, n_3 and d.

> **Note**
>
> This is not really surprising as the Cartesian equation of a plane can be scaled through by any non-zero constant without changing the plane.
>
> For example, $3x - y + 2z + 5 = 0$ and $6x - 2y + 4z + 10 = 0$ are the same plane.

You could solve to get n_1, n_2 and n_3 in terms of d:

$$n_1 = -\frac{11}{52}d, \ n_2 = \frac{7}{52}d, \ n_3 = -\frac{1}{52}d$$

So a possible form for the Cartesian equation of the plane is:

$$11x - 7y + z - 52 = 0$$

Alternatively, the vectors $\overrightarrow{AB} = \begin{pmatrix} -2 \\ -4 \\ -6 \end{pmatrix}$ and $\overrightarrow{AC} = \begin{pmatrix} 6 \\ 9 \\ -3 \end{pmatrix}$, for example, are

two vectors that lie within the plane so any vector that is perpendicular to both of these is a suitable normal vector.

$$\begin{pmatrix} 11 \\ -7 \\ 1 \end{pmatrix} \cdot \begin{pmatrix} -2 \\ -4 \\ -6 \end{pmatrix} = -22 + 28 - 6 = 0 \text{ and } \begin{pmatrix} 11 \\ -7 \\ 1 \end{pmatrix} \cdot \begin{pmatrix} 6 \\ 9 \\ -3 \end{pmatrix} = 66 - 63 - 3 = 0$$

So $\begin{pmatrix} 11 \\ -7 \\ 1 \end{pmatrix}$ is normal to the plane and, using A as a known point in the plane,

the vector equation of the plane is:

$$\begin{pmatrix} 11 \\ -7 \\ 1 \end{pmatrix} \cdot \begin{pmatrix} x - 3 \\ y + 2 \\ z - 5 \end{pmatrix} = 0 \text{ or } \begin{pmatrix} 11 \\ -7 \\ 1 \end{pmatrix} \cdot \begin{pmatrix} x \\ y \\ z \end{pmatrix} - 52 = 0$$

Another form for the equation of a plane

When you found the vector equation of a line you first moved from the origin onto the line and then travelled along the line.

$$\overrightarrow{OP} = \overrightarrow{OA} + \lambda \overrightarrow{AB} \qquad \text{or} \qquad \mathbf{r} = \mathbf{a} + \lambda \mathbf{b}$$

where $\mathbf{a}$ is the position vector of a point on the line and $\mathbf{b}$ is a vector in the direction of the line.

The same idea can be used to get a different way to write the vector equation of the plane:

$$\overrightarrow{OP} = \overrightarrow{OA} + \lambda \overrightarrow{AB} + \mu \overrightarrow{AC} \qquad \text{or} \qquad \mathbf{r} = \mathbf{a} + \lambda \mathbf{b} + \mu \mathbf{c}$$

where $\mathbf{a}$ is the position vector of a point on the plane and $\mathbf{b}$ and $\mathbf{c}$ are two vectors in the plane.

Move from the origin onto the plane and then move around in the plane by travelling some distance in one direction and some distance in another direction.

The plane through $A = (3, -2, 5)$, $B = (1, -6, -1)$ and $C = (9, 7, 2)$, with

$$\overrightarrow{AB} = \begin{pmatrix} -2 \\ -4 \\ -6 \end{pmatrix} \text{ and } \overrightarrow{AC} = \begin{pmatrix} 6 \\ 9 \\ -3 \end{pmatrix} \text{ as two vectors in the plane, can be written as:}$$

$$\mathbf{r} = \begin{pmatrix} 3 \\ -2 \\ 5 \end{pmatrix} + \lambda \begin{pmatrix} -2 \\ -4 \\ -6 \end{pmatrix} + \mu \begin{pmatrix} 6 \\ 9 \\ -3 \end{pmatrix}$$

> **Discussion point**
>
> → Can you use any vectors in the plane for $\mathbf{b}$ and $\mathbf{c}$?

Example 8.13

(i) Find, in the form $\mathbf{r} = \mathbf{a} + \lambda\mathbf{b} + \mu\mathbf{c}$, the vector equation of the plane through A, B and C where $\overrightarrow{OA} = \begin{pmatrix} 3 \\ 2 \\ 1 \end{pmatrix}$, $\overrightarrow{AB} = \begin{pmatrix} 1 \\ 3 \\ 1 \end{pmatrix}$ and $\overrightarrow{AC} = \begin{pmatrix} 0 \\ 2 \\ 1 \end{pmatrix}$.

(ii) Show that the point $(8, 13, 4)$ lies in the plane.
$P = (x, y, z)$ is a point in the plane.

(iii) Write down expressions for x, y and z in terms of λ and μ. Eliminate λ and μ to find a Cartesian form of the plane.

Solution

(i) $\mathbf{r} = \begin{pmatrix} 3 \\ 2 \\ 1 \end{pmatrix} + \lambda \begin{pmatrix} 1 \\ 3 \\ 1 \end{pmatrix} + \mu \begin{pmatrix} 0 \\ 2 \\ 1 \end{pmatrix}$

(ii) $\begin{pmatrix} 8 \\ 13 \\ 4 \end{pmatrix} = \begin{pmatrix} 3 \\ 2 \\ 1 \end{pmatrix} + \lambda \begin{pmatrix} 1 \\ 3 \\ 1 \end{pmatrix} + \mu \begin{pmatrix} 0 \\ 2 \\ 1 \end{pmatrix} \Rightarrow \begin{matrix} 8 = 3 + \lambda \\ 13 = 2 + 3\lambda + 2\mu \\ 4 = 1 + \lambda + \mu \end{matrix}$

The first equation gives $\lambda = 5$ and substituting this into the second equation gives $\mu = -2$.

These values are consistent with the third equation.

> **Note**
>
> You must check that the values of λ and μ are consistent with all three equations.

(iii) $x = 3 + \lambda$, $y = 2 + 3\lambda + 2\mu$ and $z = 1 + \lambda + \mu$

So, for example, $\lambda = x - 3$ and $\mu = z - 1 - \lambda = z - 1 - x + 3 = z - z + 2$

Then substitute into the expression for y to give:
$y = 2 + 3(x - 3) + 2(z - x + 2) = 2 + 3x - 9 + 2z - 2x + 4 = x + 2z - 3$

So,
$x - y + 2z - 3 = 0$.

The angle between planes

The angle between two planes can be found by using the scalar product. As Figures 8.19 and 8.20 show, the angle between planes π_1 and π_2 is the same as the angle between their normals, $\mathbf{n}_1$ and $\mathbf{n}_2$.

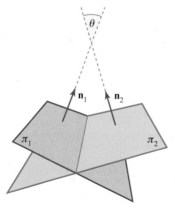

Figure 8.19

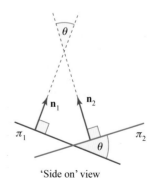

'Side on' view

Figure 8.20

Example 8.14

Find, to 1 decimal place, the acute angle between the planes
$\pi_1 : 2x + 3y + 5z = 8$ and $\pi_2 : 5x + y - 4z = 12$.

Solution

The planes have normals $\mathbf{n}_1 = \begin{pmatrix} 2 \\ 3 \\ 5 \end{pmatrix}$ and $\mathbf{n}_2 = \begin{pmatrix} 5 \\ 1 \\ -4 \end{pmatrix}$ so the angle

between the planes is given by $\mathbf{n}_1 . \mathbf{n}_2 = |\mathbf{n}_1||\mathbf{n}_2|\cos\theta$

$$\Rightarrow \begin{pmatrix} 2 \\ 3 \\ 5 \end{pmatrix} . \begin{pmatrix} 5 \\ 1 \\ -4 \end{pmatrix} = \sqrt{4 + 9 + 25}\sqrt{25 + 1 + 16}\cos\theta$$

$$\Rightarrow 10 + 3 - 20 = \sqrt{38}\sqrt{42}\cos\theta$$

$$\Rightarrow \theta = 100.1°$$

So the acute angle between the planes is 79.9°.

Exercise 8.3

① A plane has equation $5x - 3y + 2z + 1 = 0$.

(i) Write down the normal vector to this plane.

(ii) Show that the point $(1, 4, 3)$ lies on the plane.

② Find, in vector form, the equation of the planes which contain the point with position vector $\mathbf{a}$ and are perpendicular to the vector $\mathbf{n}$.

(i) $\mathbf{a} = 3\mathbf{i} + 5\mathbf{j} - 2\mathbf{k}$ $\mathbf{n} = \mathbf{i} + \mathbf{j} + \mathbf{k}$

(ii) $\mathbf{a} = -3\mathbf{i} + 2\mathbf{j} + \mathbf{k}$ $\mathbf{n} = \mathbf{i} + \mathbf{j} + \mathbf{k}$

(iii) $\mathbf{a} = 3\mathbf{i} + 5\mathbf{j} - 2\mathbf{k}$ $\mathbf{n} = -\mathbf{i} - \mathbf{j} - \mathbf{k}$

(iv) $\mathbf{a} = 2\mathbf{i} + 7\mathbf{j} - \mathbf{k}$ $\mathbf{n} = 2\mathbf{i} + 2\mathbf{j} + 2\mathbf{k}$

③ Find the Cartesian equation of the planes in question 2.

Comment on your answers.

④ (i) Find, in the form $\mathbf{r} = \mathbf{a} + \lambda\mathbf{b} + \mu\mathbf{c}$, an equation of the plane through the points $(1, 1, 1)$, $(2, 4, 5)$ and $(-1, 2, -3)$.

(ii) Why is this form not very useful when you want to find the angle between this plane and another plane?

⑤ Find, to 1 decimal place, the smaller angle between the planes:

(i) $\mathbf{r} . \begin{pmatrix} 2 \\ 2 \\ -3 \end{pmatrix} = 4$ and $\mathbf{r} . \begin{pmatrix} 3 \\ -3 \\ -1 \end{pmatrix} = 2$

(ii) $\mathbf{r} . \begin{pmatrix} 1 \\ 2 \\ -3 \end{pmatrix} = 4$ and $\mathbf{r} . \begin{pmatrix} 3 \\ -3 \\ -1 \end{pmatrix} = 2$

(iii) $x + y - 4z = 4$ and $5x - 2y + 3z = 13$

⑥ The plane π_1 has equation $-x + 3y - 2z - 13 = 0$.

Find the Cartesian and vector equations of the plane π_2 that is parallel to π_1 and passes through the point $(3, 0, -4)$.

⑦ Find the Cartesian equation of the plane which contains the point $(0, 1, -4)$ and is parallel to the plane $\left(\mathbf{r} - (4\mathbf{i} + 2\mathbf{j} - \mathbf{k})\right) \cdot (4\mathbf{i} - 5\mathbf{j} + 6\mathbf{k}) = 0$.

⑧ The planes $x - 3y - 2z = 5$ and $k^2 x + ky + 2z = 3$ are perpendicular. Find the possible values of k.

⑨ Two sloping roof structures must be constructed at an angle of exactly $60°$. The roof structures can be modelled as planes given by the equations

$$x + 2y + 2z = 5$$

$$ax + y + z = 2$$

where a is a positive constant.

Find the exact value of a.

⑩ Find the equation of the plane π which is perpendicular to the planes

$$3x - y - z + 4 = 0$$

$$x + 2y + z + 3 = 0$$

and which passes through the point $P(4, 3, 5)$.

⑪ Show that the planes $\mathbf{r} = \begin{pmatrix} 2 \\ 0 \\ -1 \end{pmatrix} + \lambda \begin{pmatrix} -1 \\ 2 \\ 2 \end{pmatrix} + \mu \begin{pmatrix} 2 \\ 1 \\ 0 \end{pmatrix}$ and

$\mathbf{r} \cdot \begin{pmatrix} -2 \\ 4 \\ -5 \end{pmatrix} = 6$ are parallel.

⑫ The points A, B and C have coordinates $(0, -1, 2), (2, 1, 0)$ and $(5, 1, 1)$.

(i) Write down the vectors $\overrightarrow{AB}$ and $\overrightarrow{AC}$.

(ii) Show that $\overrightarrow{AB} \cdot \begin{pmatrix} 1 \\ -4 \\ -3 \end{pmatrix} = \overrightarrow{AC} \cdot \begin{pmatrix} 1 \\ -4 \\ -3 \end{pmatrix} = 0$.

(iii) Find the equation of the plane containing the points A, B and C.

⑬ (i) Show that the points $A(1, 1, 1), B(3, 0, 0)$ and $C(2, 0, 2)$ all lie in the plane $2x + 3y + z = 6$.

(ii) Show that $\overrightarrow{AB} \cdot \begin{pmatrix} 2 \\ 3 \\ 1 \end{pmatrix} = \overrightarrow{AC} \cdot \begin{pmatrix} 2 \\ 3 \\ 1 \end{pmatrix} = 0$.

(iii) The point D has coordinates $(7, 6, 2)$ and lies on a line perpendicular to the plane through one of the points A, B or C.

Through which of these points does the line pass?

⑭ Three planes have equations

$$\pi_1 : ax + 2y + z = 3$$

$$\pi_2 : x + ay + z = 4$$

$$\pi_3 : x + y + az = 5$$

Given that the angle between planes π_1 and π_2 is equal to the angle between the planes π_2 and π_3, show that a must satisfy the quartic equation:

$$5a^4 + 2a^3 - 2a^2 - 8a - 3 = 0$$

⑮ A plane π has Cartesian equation $2x - 3y + 2z + 10 = 0$.

   (i) Write down the normal vector $\mathbf{n}$ and the value of $d = -\mathbf{a}.\mathbf{n}$.

   (ii) Find a possible position vector $\mathbf{a}$ to represent a point A in the plane.

   (iii) Use your answers to parts (i) and (ii) to write down a vector equation for the plane π in the form $(\mathbf{r} - \mathbf{a}).\mathbf{n} = 0$.

⑯ Four planes are given by the equations

   $\pi_1 : 2x - 3y + 5z + 4 = 0$

   $\pi_2 : 2x + 3y + z + 4 = 0$

   $\pi_3 : 4x - 6y + 10z + 4 = 0$

   $\pi_4 : 2x - 3y + z + 4 = 0$

Determine whether each *pair* of planes is parallel, perpendicular or neither.

4 Intersection of planes

> For example, the wall and ceiling of a room meet in a straight line.

If you look around you will find objects which can be used to represent planes – walls, floors, ceilings, doors, roofs and so on. You will see that in general the intersection of two planes is a straight line.

TECHNOLOGY

If you have access to 3D graphing software, investigate the different ways in which three distinct planes can intersect in three-dimensional space.

In this section you will look at the different possibilities for how *three* planes can be arranged in three-dimensional space.

Geometrical arrangement of three planes

There are five ways in which three distinct planes π_1, π_2 and π_3 can intersect in three-dimensional space.

If two of the planes are parallel, there are two possibilities for the third:

■ It can be parallel to the other two (see Figure 8.21).

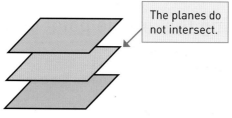

> The planes do not intersect.

Figure 8.21

■ It can cut the other two (see Figure 8.22).

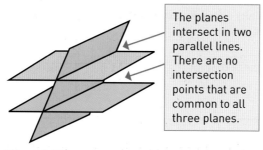

> The planes intersect in two parallel lines. There are no intersection points that are common to all three planes.

Figure 8.22

If none of the planes are parallel, there are three possibilities.

- The planes intersect in a single point (see Figure 8.23).

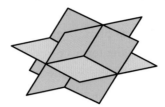

Figure 8.23

- The planes forms a **sheaf** (see Figure 8.24).

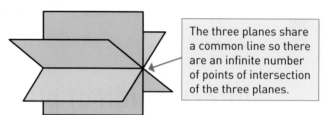

The three planes share a common line so there are an infinite number of points of intersection of the three planes.

Figure 8.24

- The planes form a **triangular prism** (see Figure 8.25).

Discussion point

➜ Think of an example from everyday life of each of these arrangements of three planes.

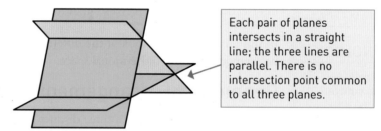

Each pair of planes intersects in a straight line; the three lines are parallel. There is no intersection point common to all three planes.

Figure 8.25

The diagrams above show that three planes intersect either in a unique point, an infinite number of points or do not have a common intersection point.

Finding the unique point of intersection of three planes

You can use 3×3 matrices to find the point of intersection of three planes that intersect in a unique point.

ACTIVITY 8.6

Make sure you remember how to find the inverse of a 3×3 matrix using your calculator.

Check that for the matrix $\mathbf{M} = \begin{pmatrix} 2 & 1 & -2 \\ 1 & 1 & 1 \\ 1 & -1 & -3 \end{pmatrix}$, $\mathbf{M}^{-1}\mathbf{M} = \mathbf{I}$.

In Chapter 7 you saw how to solve simultaneous equations using matrices. The following example shows how this relates to finding the point of intersection of three planes in three dimensions.

Example 8.15

Find the unique point of intersection of the three planes

$$2x + y - 2z = 5$$
$$x + y + z = 1$$
$$x - y - 3z = -2$$

Solution

$$\begin{pmatrix} 2 & 1 & -2 \\ 1 & 1 & 1 \\ 1 & -1 & -3 \end{pmatrix} \begin{pmatrix} x \\ y \\ z \end{pmatrix} = \begin{pmatrix} 5 \\ 1 \\ -2 \end{pmatrix}$$

> The three planes can be represented by this matrix equation.

> Solving the matrix equation will identify a point that the three planes have in common, i.e. the unique point of intersection of the three planes.

The inverse of the matrix $\mathbf{M} = \begin{pmatrix} 2 & 1 & -2 \\ 1 & 1 & 1 \\ 1 & -1 & -3 \end{pmatrix}$ is

$$\mathbf{M}^{-1} = \begin{pmatrix} -0.5 & 1.25 & 0.75 \\ 1 & -1 & -1 \\ -0.5 & 0.75 & 0.25 \end{pmatrix}.$$

> Using a calculator

> pre-multiplying both sides of the matrix equation by $\mathbf{M}^{-1}$.

$$\begin{pmatrix} -0.5 & 1.25 & 0.75 \\ 1 & -1 & -1 \\ -0.5 & 0.75 & 0.25 \end{pmatrix} \begin{pmatrix} 2 & 1 & -2 \\ 1 & 1 & 1 \\ 1 & -1 & -3 \end{pmatrix} \begin{pmatrix} x \\ y \\ z \end{pmatrix} = \begin{pmatrix} -0.5 & 1.25 & 0.75 \\ 1 & -1 & -1 \\ -0.5 & 0.75 & 0.25 \end{pmatrix} \begin{pmatrix} 5 \\ 1 \\ -2 \end{pmatrix}$$

$$\Rightarrow \begin{pmatrix} 1 & 0 & 0 \\ 0 & 1 & 0 \\ 0 & 0 & 1 \end{pmatrix} \begin{pmatrix} x \\ y \\ z \end{pmatrix} = \begin{pmatrix} -2.75 \\ 6 \\ -2.25 \end{pmatrix}$$

$$\Rightarrow \begin{pmatrix} x \\ y \\ z \end{pmatrix} = \begin{pmatrix} -2.75 \\ 6 \\ -2.25 \end{pmatrix}$$

So the planes intersect in the unique point $(-2.75, 6, -2.25)$.

Determining the other arrangements of three planes

In Example 8.15 you saw that the equations of three distinct planes can be expressed in the form

$$\mathbf{M} \begin{pmatrix} x \\ y \\ z \end{pmatrix} = \begin{pmatrix} d_1 \\ d_2 \\ d_3 \end{pmatrix}.$$

If **M** is non-singular, the planes intersect in a unique point. If **M** is singular, the planes must be arranged in one of the other four possible arrangements:

■ three parallel planes

■ two parallel planes that are cut by the third to form two parallel lines

■ a sheaf of planes that intersect in a common line

■ a prism of planes in which each pair of planes meets in a straight line but there are no common points of intersection between the three planes.

One of these cases is covered in the following example.

Example 8.16

Three planes have equations

$$2x - 5y + 3z - 2 = 0$$
$$x - y + z - 3 = 0$$
$$4x - 10y + 6z - 7 = 0$$

(i) Express the equations of the planes in the matrix form

$$\mathbf{M}\begin{pmatrix} x \\ y \\ z \end{pmatrix} = \begin{pmatrix} d_1 \\ d_2 \\ d_3 \end{pmatrix}.$$

(ii) Using your calculator, find det **M** and comment on your answer.

(iii) By comparing the rows of the matrix **M**, determine the arrangement of the three planes.

Solution

(i) The planes can be arranged in the matrix form

$$\mathbf{M}\begin{pmatrix} x \\ y \\ z \end{pmatrix} = \begin{pmatrix} 2 \\ 3 \\ 7 \end{pmatrix}$$

> Notice that the constant terms have been moved to the right hand side of each equation.

where $\mathbf{M} = \begin{pmatrix} 2 & -5 & 3 \\ 1 & -1 & 1 \\ 4 & -10 & 6 \end{pmatrix}$.

(ii) det **M** = 0

(iii) The third row is a multiple of the first row, therefore the first and third planes are parallel. The second plane is not parallel and so must cut the other two to form two parallel straight lines.

> **TECHNOLOGY**
>
> Use 3D graphing software to draw the three planes in Example 8.16.

In Example 8.16, you could see quite easily that two of the planes are parallel, but the third is not, by comparing the coefficients of x, y and z. In the same way, you would be able to identify three parallel planes.

If none of the planes are parallel, and the determinant of the matrix is zero, then the planes form either a sheaf of planes or a triangular prism. If they form a sheaf of planes, then equations are **consistent:** there are an infinite number of solutions. If they form a triangular prism, then the equations are **inconsistent:** there are no points which satisfy all three equations.

Sometimes you may have additional information which will help you to decide which arrangement you have. Otherwise, you can try to solve the equations simultaneously to find out whether the equations are consistent or inconsistent. The example below shows how this can be done.

Example 8.17

Three planes have equations

$$3x + 2y - z = 1 \quad \text{①}$$
$$x + 2y + z = 3 \quad \text{②}$$
$$x + y = 2 \qquad \text{③}$$

(i) Show that the three planes do not have a unique point of intersection.

(ii) Describe the geometrical arrangement of the three planes.

Solution

(i)
$$\begin{pmatrix} 3 & 2 & -1 \\ 1 & 2 & 1 \\ 1 & 1 & 0 \end{pmatrix} \begin{pmatrix} x \\ y \\ z \end{pmatrix} = \begin{pmatrix} 1 \\ 3 \\ 2 \end{pmatrix}$$

Using a calculator, $\det \begin{pmatrix} 3 & 2 & -1 \\ 1 & 2 & 1 \\ 1 & 1 & 0 \end{pmatrix} = 0$ so there is no unique point of

intersection.

(ii) First check if any of the planes are parallel.

The normal vectors to the three planes are all different, so none of the planes are parallel. This rules out the two cases: 3 parallel planes (Figure 8.26) and 2 parallel planes with one crossing them (Figure 8.27).

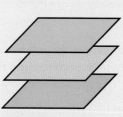

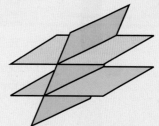

Figure 8.26   Figure 8.27

That leaves the two cases of a sheaf of planes where they all meet in the same line

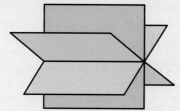

Figure 8.28

or a triangular prism.

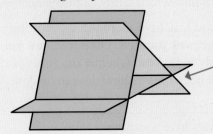

Each pair of planes intersects in a straight line; the three lines are parallel. There is no intersection point common to all three planes.

Figure 8.29

Now see if the planes meet in a single line.
Adding equations ① and ② gives $4x + 4y = 4$

$$\Rightarrow x + y = 1$$

Equation ③ is $x + y = 2$.

The value of $x + y$ cannot be both 1 and 2, so the equations are inconsistent.

So there are no points which satisfy all three equations.
Therefore the planes form a triangular prism and not a sheaf.

Exercise 8.4

① (i) Find the inverse of the matrix $\mathbf{M} = \begin{pmatrix} 2 & 1 & 3 \\ 3 & -1 & -2 \\ 1 & -2 & -1 \end{pmatrix}$.

(ii) Use your answer to part (i) to find the point of intersection of the planes

$$2x + y + 3z = 20$$
$$3x - y - 2z = 10$$
$$x - 2y - z = 30$$

② Using the same method as in question 1, find the unique point of intersection of the three planes

$$4x - 3y - 2z = 2$$
$$x + 2y + 2z = 5$$
$$3x - 3y - 2z = 3$$

③ Determine whether or not the following sets of three planes intersect in a unique point and, where possible, find the point of intersection.

(i) $x - y - 2z - 5 = 0$
$2x + y + 6z + 12 = 0$
$2x + 4y + 6z + 3 = 0$

(ii) $x + y + z - 4 = 0$
$2x + 3y - 4z - 3 = 0$
$5x + 8y - 13z - 8 = 0$

(iii) $x + 2y + 4z = 7$
$3x + 2y + 5z = 21$
$4x + y + 2z = 14$

(iv) $3x + y + z = 4$
$5x - y + 9z = 5$
$x - y + 4z = -1$

④ Three planes are given by the equations

$$-x + y + z = -1$$
$$2x + y + z = 6$$
$$x + y + z = 4$$

(i) Write the equations in the form

$$\mathbf{M}\begin{pmatrix} x \\ y \\ z \end{pmatrix} = \begin{pmatrix} d_1 \\ d_2 \\ d_3 \end{pmatrix}.$$

By comparing the rows of the matrix $\mathbf{M}$ and calculating $\det \mathbf{M}$ determine which arrangements of these planes in three dimensions are possible.

(ii) The point P $(2, 3, -1)$ is known to lie on at least one of the three planes. By working out on which planes the point P lies, determine the arrangement of the three planes.

(iii) By changing the constant term in one of the plane equations show that a different arrangement of the planes can be obtained.

⑤ Three planes are given by the equations

$$x + 2y - z = 6$$
$$2x + 4y + z = 5$$
$$3x + 6y - 3z = 8$$

Write the equations in the form

$$\mathbf{M}\begin{pmatrix} x \\ y \\ z \end{pmatrix} = \begin{pmatrix} d_1 \\ d_2 \\ d_3 \end{pmatrix}.$$

Determine the arrangement of the planes in three dimensions.

⑥ The three planes

$$x - y + 2z = k$$
$$3x - y - z = 0$$
$$2x - y + z = m$$

are known to intersect at the point $(-12, -29, -7)$.

Determine the values of k and m.

⑦ Three planes are given by the equations

$$3x + 4y + z = 5$$
$$2x - y - z = 4$$
$$5x + 14y + 5z = 7$$

(i) Write the equations in the form

$$\mathbf{M}\begin{pmatrix} x \\ y \\ z \end{pmatrix} = \begin{pmatrix} d_1 \\ d_2 \\ d_3 \end{pmatrix}.$$

By comparing the rows of the matrix $\mathbf{M}$ and calculating $\det \mathbf{M}$ determine which arrangements of the planes in three dimensions are possible.

(ii) The point P$(3, -2, 4)$ is known to lie on at least one of the three planes.

By working out on which planes the point P lies, determine the arrangement of the three planes.

⑧ The equations of three planes are
$$kx + my + nz = -6$$
$$2x - y - 2z = -9$$
$$3x + y - z = -2$$

(i) Determine the arrangement of the planes in three dimensions when $k = 1$, $m = -1$, $n = 1$, providing as much detail in your solution as possible.

(ii) State values for k, m and n which would produce an arrangement of two distinct parallel planes cut by the third plane.

(iii) Explain the arrangement of the planes in the case where $k = 9$, $m = 3$ and $n = -3$. State how this case differs from the arrangement in part (ii).

⑨ Two planes in three dimensions are said to be *coincident* if one lies on top of the other, i.e. they are exactly the same plane. Coincident planes are not distinct.

Given *any* three planes, list the ways in which they be arranged in three dimensions.

How many different possible arrangements are there in total?

5 Lines and planes

The point of intersection of a line and a plane

The next example shows how you can find the point at which a line intersects a plane.

Example 8.18

Find the point of intersection of the line $\mathbf{r} = \begin{pmatrix} 2 \\ 3 \\ 4 \end{pmatrix} + \lambda \begin{pmatrix} 1 \\ 2 \\ -1 \end{pmatrix}$ and the plane $5x + y - z = 1$.

Solution

The line is $\mathbf{r} = \begin{pmatrix} x \\ y \\ z \end{pmatrix} = \begin{pmatrix} 2 \\ 3 \\ 4 \end{pmatrix} + \lambda \begin{pmatrix} 1 \\ 2 \\ -1 \end{pmatrix}$ and so for any point on the line

$$x = 2 + \lambda \qquad y = 3 + 2\lambda \qquad z = 4 - \lambda$$

Substituting these into the equation of the plane $5x + y - z = 1$ gives

$$5(2 + \lambda) + (3 + 2\lambda) - (4 - \lambda) = 1$$
$$8\lambda = -8$$
$$\lambda = -1$$

Substituting $\lambda = -1$ into the equation of the line gives

$$\mathbf{r} = \begin{pmatrix} x \\ y \\ z \end{pmatrix} = \begin{pmatrix} 2 \\ 3 \\ 4 \end{pmatrix} - \begin{pmatrix} 1 \\ 2 \\ -1 \end{pmatrix} = \begin{pmatrix} 1 \\ 1 \\ 5 \end{pmatrix}$$

So the point of intersection is $(1, 1, 5)$.

The angle between a line and a plane

If they are not perpendicular, the acute angle between a line and a plane is the acute angle θ between the line and its **orthogonal projection** onto the plane, shown by the dotted line AB in Figure 8.30.

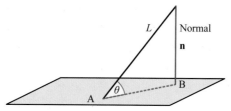

Figure 8.30

You can find the angle θ by first finding the angle α between the direction vector **d** of the straight line L and a normal vector **n** to the plane, as shown in Figure 8.31.

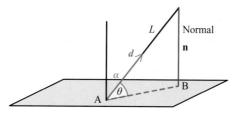

Figure 8.31

The angle θ can then be found by calculating $90 - \alpha$.

This method is illustrated in the following example.

Example 8.19

Find the angle between the line $\mathbf{r} = \begin{pmatrix} 2 \\ 0 \\ -3 \end{pmatrix} + \lambda \begin{pmatrix} 1 \\ 3 \\ 2 \end{pmatrix}$ and the plane

$3x - y + z = 4$.

Solution

For this line, the direction vector $\mathbf{d} = \begin{pmatrix} 1 \\ 3 \\ 2 \end{pmatrix}$ and a normal to the plane is

$\mathbf{n} = \begin{pmatrix} 3 \\ -1 \\ 1 \end{pmatrix}$.

The angle α between the normal vector and the direction vector satisfies

$\mathbf{d} . \mathbf{n} = |\mathbf{d}||\mathbf{n}| \cos \alpha$

$\Rightarrow \begin{pmatrix} 1 \\ 3 \\ 2 \end{pmatrix} . \begin{pmatrix} 3 \\ -1 \\ 1 \end{pmatrix} = \sqrt{14}\sqrt{11} \cos \alpha$

$\Rightarrow \cos \alpha = \dfrac{2}{\sqrt{14}\sqrt{11}}$

$\Rightarrow \alpha = 80.7°$

So the angle between the line and the plane $\theta = 90 - 80.7 = 9.3°$

① Show that the point of intersection of the line $\mathbf{r} = \begin{pmatrix} 1 \\ 3 \\ 0 \end{pmatrix} + \lambda \begin{pmatrix} 2 \\ -1 \\ 4 \end{pmatrix}$ and

the plane $2x + y + z = 26$ is $(7, 0, 12)$.

② For each of the following, find the point of intersection of the line and the plane. Find also the angle between the line and the plane.

(i) $x + 2y + 3z = 11$ $\quad \mathbf{r} = \begin{pmatrix} 1 \\ 2 \\ 4 \end{pmatrix} + \lambda \begin{pmatrix} 1 \\ 1 \\ 1 \end{pmatrix}$

(ii) $2x + 3y - 4z = 1$ $\quad \dfrac{x + 2}{3} = \dfrac{y + 3}{4} = \dfrac{z + 4}{5}$

(iii) $3x - 2y - z = 14$ $\quad \mathbf{r} = \begin{pmatrix} 8 \\ 4 \\ 2 \end{pmatrix} + \lambda \begin{pmatrix} 1 \\ 2 \\ 1 \end{pmatrix}$

(iv) $x + y + z = 0$ $\quad \mathbf{r} = \lambda \begin{pmatrix} 1 \\ 1 \\ 2 \end{pmatrix}$

③ (i) Find an equation of the line L passing through A(4, 1, 3) and B(6, 4, 8).

(ii) Find the point of intersection of L with the plane $x + 2y - z + 3 = 0$.

(iii) Find the angle between the line L and the plane.

④ (i) Find an equation of the line through $(13, 5, 0)$ parallel to the line

$$\mathbf{r} = \begin{pmatrix} 2 \\ -1 \\ 4 \end{pmatrix} + \lambda \begin{pmatrix} 3 \\ 1 \\ -2 \end{pmatrix}$$

(ii) Where does this line meet the plane $3x + y - 2z = 2$?

(iii) How far is the point of intersection from $(13, 5, 0)$?

⑤ A plane passes through the points A(2, 3, -1), B(4, 0, 1) and C(-3, 5, -2).

(i) Find the vectors $\overrightarrow{AB}$ and $\overrightarrow{AC}$.

Hence confirm that the vector $\begin{pmatrix} 1 \\ 8 \\ 11 \end{pmatrix}$ is a normal to the plane and

find the equation of the plane.

(ii) Find the points of intersection, P and Q, of the lines

$$L_1: \begin{pmatrix} 4 \\ -1 \\ 3 \end{pmatrix} + \lambda \begin{pmatrix} 2 \\ 0 \\ 1 \end{pmatrix}$$

$$L_2: \begin{pmatrix} 7 \\ -2 \\ 1 \end{pmatrix} + \mu \begin{pmatrix} -1 \\ 1 \\ 3 \end{pmatrix}$$

with the plane.

(iii) Determine the point of intersection R of the lines L_1 and L_2.

(iv) Find the angle between the vectors $\overrightarrow{PR}$ and $\overrightarrow{PQ}$.

(v) Find the area of the triangle PQR.

⑥ A laser beam ABC is fired from the point A(1, 2, 4) and is reflected at B off the plane with equation $x + 2y - 3z = 0$, as shown in Figure 8.32.

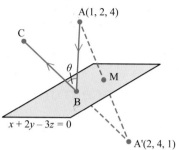

Figure 8.32

A′ is the point (2, 4, 1) and M is the midpoint of AA′.

(i) Show that AA′ is perpendicular to the plane $x + 2y - 3z = 0$ and that M lies in the plane.

The vector equation of the line AB is $\mathbf{r} = \begin{pmatrix} 1 \\ 2 \\ 4 \end{pmatrix} + \lambda \begin{pmatrix} 1 \\ -1 \\ 2 \end{pmatrix}$

(ii) Find the coordinates of B and a vector equation of the line A′B.

(iii) Given that A′BC is a straight line, find the angle θ.

⑦ Figure 8.33 shows the tetrahedron ABCD. The coordinates of the vertices are A(−3, 0, 0), B(2, 0, −2), C(0, 4, 0) and D(0, 4, 5).

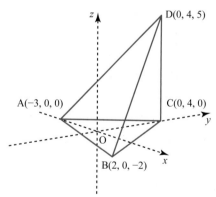

Figure 8.33

(i) Find the lengths of the edges AB and AC, and the size of the angle CAB. Hence calculate the area of the triangle ABC.

(ii) (a) Verify that $\begin{pmatrix} 4 \\ -3 \\ 10 \end{pmatrix}$ is normal to the plane ABC.

(b) Find the equation of this plane.

(iii) Write down the vector equation of the line through D that is perpendicular to the plane ABC. Find the point of intersection of this line with the plane ABC.

The area of a tetrahedron is given by $\frac{1}{3} \times$ base area $\times$ height.

(iv) Find the volume of the tetrahedron ABCD.

6 Finding distances

Sometimes you need to find the distance between points, lines and planes. In this section you will look at how to find:

- the distance from a point to a line, in two or three dimensions
- the distance from a point to a plane
- the distance between parallel or skew lines.

Finding the distance from a point to a line

Figure 8.34 shows building works at an airport that require the use of a crane near the end of the runway. How far is it from the top of the crane to the flight path of the aeroplane?

plane
taking off

runway

crane

Figure 8.34

To answer this question you need to know the flight path and the position of the top of the crane.

Working in metres, suppose the position of the top of the crane is at P(70, 30, 22) and the aeroplanes take off along the line l: $\mathbf{r} = \begin{pmatrix} -10 \\ 20 \\ 2 \end{pmatrix} + \lambda \begin{pmatrix} 5 \\ 4 \\ 3 \end{pmatrix}$ as illustrated in Figure 8.35.

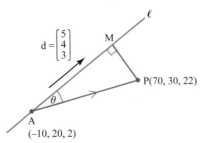

$\mathbf{d} = \begin{pmatrix} 5 \\ 4 \\ 3 \end{pmatrix}$

M

ℓ

P(70, 30, 22)

θ

A
(−10, 20, 2)

Figure 8.35

The shortest distance from P to the straight line l is measured along the line which is perpendicular to l. It is the distance PM in Figure 8.35.

The following example shows how you can find this distance.

Example 8.20

8

Chapter 8 Vectors and 3D space

The line l has equation $\mathbf{r} = \begin{pmatrix} -10 \\ 20 \\ 2 \end{pmatrix} + \lambda \begin{pmatrix} 5 \\ 4 \\ 3 \end{pmatrix}$ and the point P has coordinates $(70, 30, 22)$.

The point M is the point on l that is closest to P.

(i)   Express the position vector $\mathbf{m}$ of point M in terms of the parameter λ.

Hence find an expression for the vector $\overrightarrow{PM}$ in terms of the parameter λ.

(ii) By finding the scalar product of the vector $\overrightarrow{PM}$ with the direction vector $\mathbf{d}$, show that $\lambda = 10$ and hence find the coordinates of point M.

(iii) Find the distance PM.

Solution

(i) $\mathbf{m} = \begin{pmatrix} -10 + 5\lambda \\ 20 + 4\lambda \\ 2 + 3\lambda \end{pmatrix}$

$$\overrightarrow{PM} = \begin{pmatrix} -10 + 5\lambda - 70 \\ 20 + 4\lambda - 30 \\ 2 + 3\lambda - 22 \end{pmatrix} = \begin{pmatrix} -80 + 5\lambda \\ -10 + 4\lambda \\ -20 + 3\lambda \end{pmatrix}$$

(ii) $\overrightarrow{PM} \cdot \mathbf{d} = 0$ ←

> Since M is the point on l closest to P, PM is perpendicular to l and so PM is perpendicular to $\mathbf{d}$.

$$\begin{pmatrix} -80 + 5\lambda \\ -10 + 4\lambda \\ -20 + 3\lambda \end{pmatrix} \cdot \begin{pmatrix} 5 \\ 4 \\ 3 \end{pmatrix} = 0$$

$$5(-80 + 5\lambda) + 4(-10 + 4\lambda) + 3(-20 + 3\lambda) = 0$$

$$-400 + 25\lambda - 40 + 16\lambda - 60 + 9\lambda = 0$$

$$50\lambda = 500$$

$$\lambda = 10$$

$$\mathbf{m} = \begin{pmatrix} -10 + 5\lambda \\ 20 + 4\lambda \\ 2 + 3\lambda \end{pmatrix} = \begin{pmatrix} 40 \\ 60 \\ 32 \end{pmatrix}$$

(iii) $\overrightarrow{PM} = \begin{pmatrix} -30 \\ 30 \\ 10 \end{pmatrix}$

$$|\overrightarrow{PM}| = 10\sqrt{(-3)^2 + 3^2 + 1^2} = 10\sqrt{19}$$

The distance from a point to a plane

The distance from the point $P(x_1, y_1, z_1)$ to the plane π with equation $ax + by + cz + d = 0$ is PM, where M is the foot of the perpendicular from P to the plane (see Figure 8.35).

Notice that since PM is normal to the plane, it is parallel to the vector $\mathbf{n} = a\mathbf{i} + b\mathbf{j} + c\mathbf{k}$.

Take any point, other than M, on the plane and call it R, with position vector **r**.

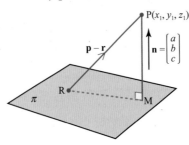

Figure 8.36

If the angle between the vectors **p** − **r** and **n** is acute (as shown in Figure 8.36):

$$PM = RP\cos R\hat{P}M = \overrightarrow{RP} \cdot \hat{\mathbf{n}}$$

Using the scalar product
$$\mathbf{a} \cdot \mathbf{b} = |\mathbf{a}||\mathbf{b}|\cos\theta$$

$$= (\mathbf{p} - \mathbf{r}) \cdot \hat{\mathbf{n}}$$

If the angle between **p** − **r** and **n** is obtuse, $\cos R\hat{P}M$ is negative and
$$PM = -(\mathbf{p} - \mathbf{r}) \cdot \hat{\mathbf{n}}$$

Now you want to choose coordinates for the point R that will keep your

$\left(0, 0, -\dfrac{d}{c}\right)$ lies on the plane $ax + by + cz = d$.

→ working simple. A suitable point is $\left(0, 0, -\dfrac{d}{c}\right)$. For this point, $\mathbf{r} = \begin{pmatrix} 0 \\ 0 \\ -\dfrac{d}{c} \end{pmatrix}$.

Then $(\mathbf{p} - \mathbf{r}) \cdot \mathbf{n} = \begin{pmatrix} x_1 \\ y_1 \\ z_1 + \dfrac{d}{c} \end{pmatrix} \cdot \begin{pmatrix} a \\ b \\ c \end{pmatrix} = ax_1 + by_1 + cz_1 + d$

and $PM = |(\mathbf{p} - \mathbf{r}) \cdot \hat{\mathbf{n}}|$

$$= \left| \frac{(\mathbf{p} - \mathbf{r}) \cdot \mathbf{n}}{|\mathbf{n}|} \right| = \frac{|ax_1 + by_1 + cz_1 + d|}{\sqrt{a^2 + b^2 + c^2}}$$

Example 8.21

Find the shortest distance from the point $(2, 4, -2)$ to the plane $6x - y - 3z + 1 = 0$.

Solution

The shortest distance from the point (x_1, y_1, z_1) to the plane $ax + by + cz + d = 0$ is

$$\frac{|ax_1 + by_1 + cz_1 + d|}{\sqrt{a^2 + b^2 + c^2}}$$

In this case, $x_1 = 2$, $y_1 = 4$, $z_1 = -2$ and $a = 6$, $b = -1$, $c = -3$, $d = 1$

so the shortest distance from the point to the plane is

$$\frac{|ax_1 + by_1 + cz_1 + d|}{\sqrt{a^2 + b^2 + c^2}}$$

$$= \frac{(6 \times 2) + (-1 \times 4) + (-3 \times -2) + 1}{\sqrt{6^2 + (-1)^2 + (-3)^2}}$$

$$= \frac{15}{\sqrt{46}} \approx 2.21$$

Finding the distance between two parallel lines

The distance between two parallel lines l_1 and l_2 is measured along a line PQ which is perpendicular to both l_1 and l_2, as shown in Figure 8.37.

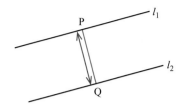

Figure 8.37

You can find this distance by simply choosing a point P on l_1, say, and then finding the shortest distance from P to the line l_2.

Example 8.22

Two straight lines in three dimensions are given by the equations:

$$l_1: \begin{pmatrix} 2 \\ -3 \\ 0 \end{pmatrix} + \lambda \begin{pmatrix} 1 \\ -3 \\ 2 \end{pmatrix} \quad \text{and} \quad l_2: \begin{pmatrix} 4 \\ 2 \\ 1 \end{pmatrix} + \mu \begin{pmatrix} -2 \\ 6 \\ -4 \end{pmatrix}$$

(i) Show that the two lines are parallel.

(ii) Find the shortest distance between the two lines.

Solution

(i) The direction vectors of the two lines are $\mathbf{d}_1 = \begin{pmatrix} 1 \\ -3 \\ 2 \end{pmatrix}$ and $\mathbf{d}_2 = \begin{pmatrix} -2 \\ 6 \\ -4 \end{pmatrix}$.

Since $\mathbf{d}_2 = -2\mathbf{d}_1$ the two lines are parallel.

> You could use any value for μ.

(ii) Choose a point P on l_2 by setting $\mu = 0$ which gives $\mathbf{p} = \begin{pmatrix} 4 \\ 2 \\ 1 \end{pmatrix}$.

Let the point Q be the point on l_1 that is closest to P.

$$\mathbf{q} = \begin{pmatrix} 2 + \lambda \\ -3 - 3\lambda \\ 2\lambda \end{pmatrix}$$

$$\overrightarrow{PQ} = \begin{pmatrix} 2 + \lambda \\ -3 - 3\lambda \\ 2\lambda \end{pmatrix} - \begin{pmatrix} 4 \\ 2 \\ 1 \end{pmatrix} = \begin{pmatrix} -2 + \lambda \\ -5 - 3\lambda \\ 2\lambda - 1 \end{pmatrix}$$

$\overrightarrow{PQ}$ is perpendicular to $\mathbf{d}_1 = \begin{pmatrix} 1 \\ -3 \\ 2 \end{pmatrix}$, so

$$\begin{pmatrix} -2 + \lambda \\ -5 - 3\lambda \\ 2\lambda - 1 \end{pmatrix} \cdot \begin{pmatrix} 1 \\ -3 \\ 2 \end{pmatrix} = 0$$

$$-2 + \lambda - 3(-5 - 3\lambda) + 2(2\lambda - 1) = 0$$

$$14\lambda = -11$$

$$\lambda = -\frac{11}{14}$$

$$\overrightarrow{PQ} = \begin{pmatrix} -2 - \frac{11}{14} \\ -5 + 3 \times \frac{11}{14} \\ 2 \times \left(-\frac{11}{14}\right) - 1 \end{pmatrix} = \begin{pmatrix} -\frac{39}{14} \\ -\frac{37}{14} \\ -\frac{36}{14} \end{pmatrix}$$

The shortest distance is $\left|\overrightarrow{PQ}\right| = \dfrac{\sqrt{299}}{\sqrt{14}} \approx 4.62$ units.

Finding the distance between skew lines

Two lines are skew if they do not intersect and are not parallel.

Figure 8.38 shows two skew lines l_1 and l_2. The shortest distance between the two lines is measured along a line that is perpendicular to both l_1 and l_2.

The common perpendicular of l_1 and l_2 is the perpendicular from l_1 that passes through the point Q, the point of intersection of l_2 and l'_1.

Drop perpendiculars from the points on l_1 to π to form l'_1, which is the projection of l_1 on π.

π is the plane containing l'_1 parallel to l_1.

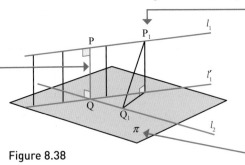

Figure 8.38

ACTIVITY 8.7

Explain why PQ is shorter than any other line, such as P_1Q_1 joining lines l_1 and l_2.

Figure 8.39 shows the lines l_1 and l_2 and two parallel planes. l_1 and l_2 have equations $\mathbf{r} = \mathbf{a}_1 + \lambda\mathbf{d}_1$ and $\mathbf{r} = \mathbf{a}_2 + \mu\mathbf{d}_2$ respectively. A_1 and A_2 are points on the lines l_1 and l_2 with position vectors $\mathbf{a}_1$ and $\mathbf{a}_2$ respectively.

π_1 contains l_1 and is parallel to l_2

π_2 contains l_2 and is parallel to l_1

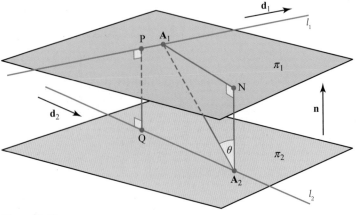

Figure 8.39

Then PQ, the common perpendicular of l_1 and l_2 has the same length as any other perpendicular between the planes, such as A_2N. If angle $A_1A_2N = \theta$ then

$$PQ = A_2N = A_2A_1\cos\theta = \left|(\mathbf{a}_2 - \mathbf{a}_1)\cdot\hat{\mathbf{n}}\right|$$

where $\hat{\mathbf{n}}$ is a unit vector parallel to A_2N, i.e. perpendicular to both planes.

Notice that the modulus function is used to ensure a positive answer: the vector $\hat{\mathbf{n}}$ may be directed from π_1 to π_2 making $(\mathbf{a}_2 - \mathbf{a}_1)\cdot\hat{\mathbf{n}}$ negative.

You can find the vector $\hat{\mathbf{n}}$ by first solving the equations $\mathbf{n}.\mathbf{d}_1 = 0$ and $\mathbf{n}.\mathbf{d}_2 = 0$ to find a vector $\mathbf{n}$ that is perpendicular to both planes, and then finding the unit vector in the direction of $\mathbf{n}$ by dividing by $|\mathbf{n}|$.

Example 8.23

Find the shortest distance between the lines $l_1: \mathbf{r} = \begin{pmatrix} 8 \\ 9 \\ -2 \end{pmatrix} + \lambda\begin{pmatrix} 1 \\ 2 \\ -3 \end{pmatrix}$ and $l_2: \mathbf{r} = \begin{pmatrix} 6 \\ 0 \\ -2 \end{pmatrix} + \mu\begin{pmatrix} 1 \\ -1 \\ -2 \end{pmatrix}$

Solution

Line l_1 contains the point $A_1(8, 9, -2)$ and is parallel to the vector $\mathbf{d}_1 = \mathbf{i} + 2\mathbf{j} - 3\mathbf{k}$.

Line l_2 contains the point $A_2(6, 0, -2)$ and is parallel to the vector $\mathbf{d}_2 = \mathbf{i} - \mathbf{j} - 2\mathbf{k}$.

Let $\mathbf{n} = \begin{pmatrix} n_1 \\ n_2 \\ n_3 \end{pmatrix}$ be perpendicular to both planes.

So $\begin{pmatrix} n_1 \\ n_2 \\ n_3 \end{pmatrix}\cdot\begin{pmatrix} 1 \\ 2 \\ -3 \end{pmatrix} = 0 \implies n_1 + 2n_2 - 3n_3 = 0$ \quad (1)

and $\begin{pmatrix} n_1 \\ n_2 \\ n_3 \end{pmatrix}\cdot\begin{pmatrix} 1 \\ -1 \\ -2 \end{pmatrix} = 0 \implies n_1 - n_2 - 2n_3 = 0$ \quad (2)

Subtracting equation (2) from (1): $\implies 3n_2 - n_3 = 0$

$\implies n_3 = 3n_2$

Taking $n_2 = 1$ gives $n_3 = 3$ and $n_1 = 7$

so $\mathbf{n} = \begin{pmatrix} 7 \\ 1 \\ 3 \end{pmatrix}$ and $\hat{\mathbf{n}} = \frac{1}{\sqrt{59}}\begin{pmatrix} 7 \\ 1 \\ 3 \end{pmatrix}$

$$\mathbf{a}_1 - \mathbf{a}_2 = \begin{pmatrix} 8 \\ 9 \\ -2 \end{pmatrix} - \begin{pmatrix} 6 \\ 0 \\ -2 \end{pmatrix} = \begin{pmatrix} 2 \\ 9 \\ 0 \end{pmatrix}$$

So the shortest distance $= \dfrac{1}{\sqrt{59}} \begin{pmatrix} 2 \\ 9 \\ 0 \end{pmatrix} \cdot \begin{pmatrix} 7 \\ 1 \\ 3 \end{pmatrix}$

$$= \dfrac{23}{\sqrt{59}} \approx 2.99 \text{ units}$$

Exercise 8.6

① Calculate the distance from the point P to the line l.

   (i) P$(1, -2, 3)$ $l\!: \dfrac{x-1}{2} = \dfrac{y-5}{2} = \dfrac{z+1}{-1}$

   (ii) P$(2, 3, -5)$ $l\!: \mathbf{r} = \begin{pmatrix} 4 \\ 3 \\ 4 \end{pmatrix} + \lambda \begin{pmatrix} 6 \\ -7 \\ 6 \end{pmatrix}$

   (iii) P$(8, 9, 1)$ $l\!: \dfrac{x-6}{12} = \dfrac{y-5}{-9} = \dfrac{z-11}{-8}$

② Find the distance from the point P to the line l.

   (i)   P$(8, 9)$ $l\!: 3x + 4y + 5 = 0$

   (ii) P$(5, -4)$ $l\!: 6x - 3y + 3 = 0$

   (iii) P$(4, -4)$ $l\!: 8x + 15y + 11 = 0$

③ Find the distance from the point P to the plane π.

   (i)   P$(5, 4, 0)$ $\pi\!: 6x + 6y + 7z + 1 = 0$

   (ii) P$(7, 2, -2)$ $\pi\!: 12x - 9y - 8z + 3 = 0$

   (iii) P$(-4, -5, 3)$ $\pi\!: 8x + 5y - 3z - 4 = 0$

④ A line l_1 has equation $\mathbf{r} = \begin{pmatrix} 2 \\ 0 \\ -1 \end{pmatrix} + \lambda \begin{pmatrix} 1 \\ -2 \\ -1 \end{pmatrix}$.

   (i) Write down an equation of a line parallel to l_1 passing through the point $(3, 1, 0)$.

   (ii) Find the distance between these two lines.

⑤ (i) Show that the lines $\mathbf{r} = \begin{pmatrix} 1 \\ 2 \\ 4 \end{pmatrix} + \lambda \begin{pmatrix} 3 \\ 0 \\ 2 \end{pmatrix}$ and $\mathbf{r} = \begin{pmatrix} 2 \\ 1 \\ 0 \end{pmatrix} + \mu \begin{pmatrix} 1 \\ 1 \\ -1 \end{pmatrix}$ are skew.

   (ii) Find the shortest distance between these two lines.

⑥ Find the shortest distance between the lines l_1 and l_2.

   In each case, state whether the lines are skew, parallel or intersect.

   (i) $l_1\!: \dfrac{x-2}{1} = \dfrac{y-3}{2} = \dfrac{z-4}{2}$ and $l_2\!: \dfrac{x-2}{2} = \dfrac{y-9}{-2} = \dfrac{z-1}{1}$

   (ii) $l_1\!: \dfrac{x-8}{4} = \dfrac{y+2}{3} = \dfrac{z-7}{5}$ and $l_2\!: \dfrac{x-2}{2} = \dfrac{y+6}{-6} = \dfrac{z-1}{-9}$

(iii) $l_1: \mathbf{r} = \begin{pmatrix} -5 \\ 6 \\ 1 \end{pmatrix} + \lambda \begin{pmatrix} 8 \\ 6 \\ 3 \end{pmatrix}$ and $l_2: \mathbf{r} = \begin{pmatrix} 5 \\ 8 \\ 3 \end{pmatrix} + \mu \begin{pmatrix} 5 \\ 1 \\ 1 \end{pmatrix}$

(iv) $l_1: \mathbf{r} = \begin{pmatrix} 2 \\ 3 \\ -1 \end{pmatrix} + \lambda \begin{pmatrix} 1 \\ 1 \\ 2 \end{pmatrix}$ and $l_2: \mathbf{r} = \begin{pmatrix} 4 \\ 0 \\ -1 \end{pmatrix} + \lambda \begin{pmatrix} -2 \\ -2 \\ -4 \end{pmatrix}$

⑦ (i) Find the shortest distance from the point P(13, 4, 2) to the line
$l: \mathbf{r} = \begin{pmatrix} 2 \\ -8 \\ -21 \end{pmatrix} + \lambda \begin{pmatrix} 1 \\ -2 \\ 3 \end{pmatrix}$.

(ii) Find the coordinates of the point M which is the foot of the perpendicular from P to the line l.

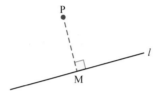

Figure 8.40

⑧ (i) Find the exact distance from the point A(2, 0, −5) to the plane $\pi: 4x − 5y + 2z + 4 = 0$.

(ii) Write down the equation of the line l through the point A that is perpendicular to the plane π.

(iii) Find the exact coordinates of the point M where the perpendicular from the point A meets the plane π.

⑨ In a school production some pieces of the stage set are held in place by steel cables. The location of points on the cables can be measured, in metres, from an origin O at the side of the stage.

Cable 1 passes through the points A(2, −3, 4) and B(1, −3, 5) while cable 2 passes through the points C(0, 3, −2) and D(2, 3, 5).

(i) Find the vector equations of the lines AB and CD and determine the shortest distance between these two cables.

One piece of the stage set, with corner at E(1, 6, −1), needs to be more firmly secured with an additional cable. It is decided that the additional cable should be attached to cable 2.

(ii) If the additional cable available is three metres long, determine whether it will be long enough to attach point E to cable 2.

⑩ The point P has coordinates (4, k, 5) where k is a constant.

The line L has equation $\mathbf{r} = \begin{pmatrix} 1 \\ 0 \\ -4 \end{pmatrix} + \lambda \begin{pmatrix} 1 \\ 2 \\ -2 \end{pmatrix}$

The line M has equation $\mathbf{r} = \begin{pmatrix} 4 \\ k \\ 5 \end{pmatrix} + \mu \begin{pmatrix} 7 \\ 3 \\ -4 \end{pmatrix}$

(i) Show that the shortest distance from the point P to the line L is $\frac{1}{3}\sqrt{5(k^2 + 12k + 117)}$.

(ii) Find, in terms of k, the shortest distance between the lines L and M.

(iii) Find the value of k for which the lines L and M intersect.

(iv) When $k = 12$, show that the distances in parts (i) and (ii) are equal. In this case, find the equation of the line which is perpendicular to, and intersects, both L and M.

⑪ The point A$(-1, 12, 5)$ lies on the plane P with equation $8x - 3y + 10z = 6$. The point B$(6, -2, 9)$ lies on the plane Q with equation $3x - 4y - 2z = 8$. The planes P and Q intersect in the line L.

(i) (a) Show that the point $(0, -2, 0)$ lies on both planes.

(b) Find a vector perpendicular to both the normal to plane P and the normal to plane Q.

(c) Write down the equation of L, the line of intersection of planes P and Q.

(ii) Find the shortest distance between L and the line AB.

The lines M and N are both parallel to L, with M passing through A and N passing through B.

(iii) Find the distance between the parallel lines M and N.

The point C has coordinates $(k, 0, 2)$ and the line AC intersects the line N at the point D.

(iv) Find the value of k and the coordinates of D.

Adapted from MEI June 2009 FP3

LEARNING OUTCOMES

When you have completed this chapter you should be able to:

➤ find the scalar product of two vectors

➤ use the scalar product to find the angle between two vectors

➤ know that two vectors are perpendicular if and only if their scalar product is zero

➤ form the equation of a line in three dimensions in vector or Cartesian form

➤ find out whether two lines in three dimensions are parallel, skew or intersect, and find the point of intersection if there is one

➤ find the angle between two lines

➤ identify a vector normal to a plane, given the equation of the plane

➤ find the equation of a plane in vector or Cartesian form

➤ find the angle between two planes

➤ know the different ways in which three distinct planes can be arranged in 3-D space

➤ understand how solving three linear simultaneous equations in three unknowns relates to finding the point of intersection of three planes in three dimensions

➤ find the point of intersection of a line and a plane

➤ find the angle between a line and a plane

➤ find the perpendicular distance from a point to a line, from a point to a plane, and between two lines.

KEY POINTS

1 In 2 dimensions, the scalar product is

$$\mathbf{a} \cdot \mathbf{b} = \begin{pmatrix} a_1 \\ a_2 \end{pmatrix} \cdot \begin{pmatrix} b_1 \\ b_2 \end{pmatrix} = a_1b_1 + a_2b_2 = |\mathbf{a}||\mathbf{b}| \cos\theta.$$

2 In 3 dimensions, the scalar product is

$$\mathbf{a} \cdot \mathbf{b} = \begin{pmatrix} a_1 \\ a_2 \\ a_3 \end{pmatrix} \cdot \begin{pmatrix} b_1 \\ b_2 \\ b_3 \end{pmatrix} = a_1b_1 + a_2b_2 + a_3b_3 = |\mathbf{a}||\mathbf{b}| \cos\theta.$$

3 The angle θ between two vectors $\mathbf{a}$ and $\mathbf{b}$ is given by

$$\cos\theta = \frac{\mathbf{a} \cdot \mathbf{b}}{|\mathbf{a}||\mathbf{b}|} \quad \text{where} \quad \begin{aligned} \mathbf{a} \cdot \mathbf{b} &= a_1b_1 + a_2b_2 \quad \text{(in two dimensions)} \\ \mathbf{a} \cdot \mathbf{b} &= a_1b_1 + a_2b_2 + a_3b_3 \quad \text{(in three dimensions)} \end{aligned}$$

4 The vector equation of the line through the point with position vector $\mathbf{a}$ and direction vector $\mathbf{b}$ is given by $\mathbf{r} = \mathbf{a} + \lambda\mathbf{b}$, where λ is a parameter.

5 The Cartesian equation of the line through the point with position vector

$$\mathbf{a} = \begin{pmatrix} a_1 \\ a_2 \\ a_3 \end{pmatrix} \text{ and direction vector } \mathbf{b} = \begin{pmatrix} b_1 \\ b_2 \\ b_3 \end{pmatrix} \text{ is given by}$$

$$\frac{x - a_1}{b_1} = \frac{y - a_2}{b_2} = \frac{z - a_3}{b_3}$$

6 Two lines are parallel if their direction vectors are multiples of one another.

7 If two lines are not parallel they may intersect in a point or they may be skew.

To find where two lines intersect, equate their x terms and equate their y terms. Solve the resulting equations to find the values of the parameters.

If these parameter values are consistent with the z terms the lines intersect in the point found by using these parameter values.

If the parameter values are inconsistent with the z terms the lines do not intersect.

8 The angle between two lines is the same as the angle between their direction vectors.

9 The Cartesian equation of the plane perpendicular to the vector

$$\mathbf{n} = \begin{pmatrix} n_1 \\ n_2 \\ n_3 \end{pmatrix} \text{ and passing through a point with position vector } \mathbf{a} \text{ is given by}$$

$n_1x + n_2y + n_3z + d = 0$ where $d = -\mathbf{a} \cdot \mathbf{n}$

10 The vector equation of the plane through the point with position vector $\mathbf{a}$ and

perpendicular to the vector $\mathbf{n} = \begin{pmatrix} n_1 \\ n_2 \\ n_3 \end{pmatrix}$ is given by $(\mathbf{r} - \mathbf{a}) \cdot \mathbf{n} = 0$ or $\mathbf{r} \cdot \mathbf{n} = k$,

where k is the value of $\mathbf{a} \cdot \mathbf{n}$

11 The vector equation of a plane can be written as

$\mathbf{r} = \mathbf{a} + \lambda\mathbf{b} + \mu\mathbf{c}$

where $\mathbf{a}$ is the position vector of a point on the plane and $\mathbf{b}$ and $\mathbf{c}$ are two (non-parallel) vectors in the plane.

12 The angle between two planes π_1 and π_2 is the same as the angle between their normals, $\mathbf{n}_1$ and $\mathbf{n}_2$. This angle can be found using the scalar product.

13 Three distinct planes in three dimensions will be arranged in one of five ways:

  ■ meet in a unique point of intersection
  ■ three parallel planes
  ■ two parallel planes that are crossed by the third to form two parallel lines
  ■ a triangular prism in which each pair of planes meets in a straight line but there are no common points of intersection between the three planes
  ■ a sheaf of planes that intersect in a common line.

14 Three distinct planes $a_1x + b_1y + c_1z = d_1$, $a_2x + b_2y + c_2z = d_2$ and

$a_3x + b_3y + c_3z = d_3$ can be expressed in the form $\mathbf{M}\begin{pmatrix} x \\ y \\ z \end{pmatrix} = \begin{pmatrix} d_1 \\ d_2 \\ d_3 \end{pmatrix}$

where $\mathbf{M} = \begin{pmatrix} a_1 & b_1 & c_1 \\ a_2 & b_2 & c_2 \\ a_3 & b_3 & c_3 \end{pmatrix}$.

If $\mathbf{M}$ is non-singular, the unique point of intersection is given by $\mathbf{M}^{-1}\begin{pmatrix} d_1 \\ d_2 \\ d_3 \end{pmatrix}$.

Otherwise, the planes meet in one of the other four possible arrangements (three parallel planes, two parallel planes that are crossed by the third, a triangular prism or a sheaf). In the case of a sheaf of planes, the equations have an infinite number of possible solutions (a line of solutions), in the other three cases the equations have no solution.

15 To find where a line intersects a plane, substitute for $\mathbf{r}$ from the line into the plane.

16 The angle between a line and a plane is found by finding the angle between the line and the normal to the plane.

17 The shortest distance from the point P, with position vector $\mathbf{p}$, to the line $\mathbf{r} = \mathbf{a} + \lambda\mathbf{b}$ is

$|(\mathbf{a} - \mathbf{p}) + \lambda\mathbf{b}|$, where $\lambda = \dfrac{(\mathbf{p} - \mathbf{a})\cdot\mathbf{b}}{\mathbf{b}\cdot\mathbf{b}}$.

18 The perpendicular distance from the point (α, β, γ) to the plane $n_1x + n_2y + n_3z + d = 0$ is

$\dfrac{|n_1\alpha + n_2\beta + n_3\gamma + d|}{\sqrt{n_1^2 + n_2^2 + n_3^2}}$

19 The shortest distance between the lines $\mathbf{r}_1 = \mathbf{a} + \lambda\mathbf{b}$ and $\mathbf{r}_2 = \mathbf{c} + \mu\mathbf{d}$ is

$\dfrac{|(\mathbf{c} - \mathbf{a})\cdot\mathbf{n}|}{|\mathbf{n}|}$, where $\mathbf{n}$ is perpendicular to $\mathbf{b}$ and $\mathbf{d}$ ($\mathbf{n}.\mathbf{b} = 0$ and $\mathbf{n}.\mathbf{d} = 0$).

FUTURE USES

■ The work on vectors is developed further in A2 Further Mathematics.

① (i) Describe the transformation represented by the matrix

$$\mathbf{A} = \begin{pmatrix} 1 & 0 \\ 0 & -1 \end{pmatrix}.$$ [1 mark]

(ii) Describe the transformation represented by the matrix

$$\mathbf{B} = \begin{pmatrix} -1 & 0 \\ 0 & 1 \end{pmatrix}.$$ [1 mark]

(iii) Determine **BA** and describe the transformation it represents. [2 marks]

(iv) Determine **(BA)**$^{-1}$. What do you notice? Explain your answer in terms of the transformation represented by **BA**. [3 marks]

MP ② $z_1 = a + bi$ and $z_2 = c + di$, where a, b, c and d are real numbers.

(i) Work out $z_1 z_2$. [2 marks]

(ii) Write down $|z_1|$ and $|z_2|$. [1 mark]

(iii) Prove that $|z_1 z_2| = |z_1||z_2|$. [4 marks]

PS ③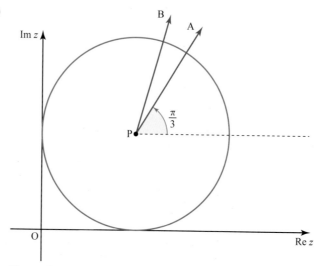

Figure 8.41

On the Argand diagram above, the point P is at $3 + 3i$.
P is the centre of a circle that touches the axes.

(i) Write down the locus of the circle in terms of z. [2 marks]

(ii) Write down the equation of the locus of points represented by the half line from P through A. [2 marks]

The sector PAB has area $\dfrac{3\pi}{8}$.

(iii) Find the equation of the locus of points represented by the half line from P through B. [4 marks]

④ The plane p contains the point $(5, 0, 4)$. The vector $\begin{pmatrix} 5 \\ -1 \\ 0 \end{pmatrix}$ is perpendicular to p.

(i) Find the equation of p in the form $ax + by + cz + d = 0$. [2 marks]

Another plane, q, has equation $4x - 3y + z - 3 = 0$.

(ii) Find the angle between p and q. [3 marks]

(iii) Show that the point $(5, 0, -17)$ lies on both planes. [2 marks]

⑤ (i) Sketch the curve $y = 1 - \cos x$ between $x = 0$ and $x = 2\pi$. [1 mark]

A solid is formed by rotating the curve $y = 1 - \cos x$ through 2π about the x-axis, between $x = 0$ and $x = 2\pi$.

(ii) Use the results $\cos^2 x = \frac{1}{2}(1 + \cos 2x)$ and $\int \cos kx \, dx = \frac{1}{k}\sin kx + c$

to calculate the volume of the solid. [4 marks]

⑥ The plane Π passes through the points $(0, 1, 2)$, $(1, 1, 2)$ and $(3, 3, 3)$.

(i) Give the equation of Π in the form $n_1 x + n_2 y + n_3 z + d = 0$. [3 marks]

The line L passes through the points $(1, 2, 3)$ and $(4, 2, 7)$

The equation of L can be written as $\begin{pmatrix} x \\ y \\ z \end{pmatrix} = \begin{pmatrix} 1 \\ 2 \\ 3 \end{pmatrix} + t\mathbf{b}$

(ii) Write down a suitable vector for $\mathbf{b}$. [1 mark]

(iii) Calculate the coordinates of the point where the line L cuts the plane Π. [3 marks]

(iv) Calculate the shortest distance from $(3, 3, 3)$ to the line L. [3 marks]

Ⓜ ⑦ A new skyscraper is built in the shape of square-based pyramid, standing on its square base. In a model of the skyscraper, its four triangular faces are parts of four planes and the ground on which it stands is the plane $z = 0$.

The summit of the skyscraper, where its four triangular faces meet, is directly above the centre of its square base and has coordinates $(5, 17, 20)$.

The faces of the skyscraper are modelled by the four planes:

$20x + z = k$

$-20x + z = l$

$-20y - z = m$

$20y - z = n$

(i) Find the values of k, l, m and n. [2 marks]

(ii) What angle does each of the skyscraper's sides make with the vertical? [4 marks]

The triangular sides of another skyscraper built in the form of a square-based pyramid are modelled as parts of these four planes:

$25x + z = 150$

$-25x + z = -100$

$25y - z = 250$

$25y + z = 300$

(iii) What are the coordinates of its summit? [3 marks]

(iv) The length of one unit in this question has not been defined. Given that this is an extremely tall skyscraper, suggest and justify an actual length for 1 unit. [3 marks]

⑧ The equations of three planes are:

$$5x - 7y + z = 80$$

$$ax - by - 2z = c$$

$$19x + 17y - 4z = -14$$

(i) State a set of values for a, b and c for which two of the planes are coincident. [3 marks]

(ii) State a set of values for a, b and c for which there are two distinct parallel planes that are cut by a third plane. [2 marks]

(iii) If $a = 1$, $b = 13$ and $c = -2$, show that the planes must meet at a single point and find the coordinates of that point. [6 marks]

An introduction to radians

Radians are an alternative way to measure angles. They relate the arc length of a sector to its angle. In Figure 1 the arc AB has been drawn so that it is equal to the length of the radius, *r*. The angle subtended at the centre of the circle is one radian.

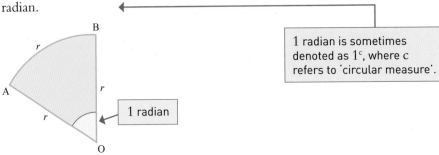

1 radian is sometimes denoted as 1ᶜ, where c refers to 'circular measure'.

Figure 1

Since an angle of 1 radian at the centre of the circle corresponds to an arc length *r* it follows that an angle of 2 radians corresponds to an arc length of 2*r* and so on. In general, an angle of θ radians corresponds to an arc length of $r\theta$, as shown in Figure 2.

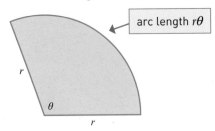

arc length $r\theta$

Figure 2

The circumference of a circle is $2\pi r$, so the angle at the centre of a full circle is $2\pi r$ radians. This is 360°.

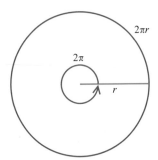

Figure 3

So $360° = 2\pi$ radians

$180° = \pi$ radians

$90° = \dfrac{\pi}{2}$ radians

$60° = \dfrac{\pi}{3}$ radians

$45° = \dfrac{\pi}{4}$ radians

$30° = \dfrac{\pi}{6}$ radians

When working in radians, angles are often stated as a fraction or multiple of π.

$360° = 2\pi^c$ and so 1 radian is equivalent to $360 \div 2\pi = 57.3°$ to one decimal place.

The fact that one radian is just under 60° can be a helpful reference point.

When a multiple of π is used the 'c' symbol is usually omitted, as it is implied that the measure is radians.

To convert degrees into radians you multiply by $\dfrac{\pi}{180}$, and to convert radians into degrees you multiply by $\dfrac{180}{\pi}$.

An introduction to radians

Example

(i)   Express in radians, giving your answers as a multiple of π:

     (a) 120°   (b) 225°   (c) 390°

(ii)   Express in radians, giving your answers to 3 significant figures:

     (a) 34°   (b) 450°   (c) 1°

(iii)   Express in degrees, giving your answers to 3 significant figures where appropriate:

     (a) $\dfrac{5\pi}{12}$ (b) $\dfrac{\pi}{24}$ (c) 3.4^c

Solution

(i)   (a) $60° = \dfrac{\pi}{3}$ radians so $120° = \dfrac{2\pi}{3}$ radians

     (b) $45° = \dfrac{\pi}{4}$ radians so $225° = 5 \times 45° = \dfrac{5\pi}{4}$

     (c) $30° = \dfrac{\pi}{6}$ radians so $390° = 360° + 30° = 2\pi + \dfrac{\pi}{6} = \dfrac{13\pi}{6}$

(ii)   (a) $34 \times \dfrac{\pi}{180} = 0.593$ radians

     (b) $450 \times \dfrac{\pi}{180} = 7.85$ radians

     (c) $1 \times \dfrac{\pi}{180} = 0.0175$ radians

(iii)   (a) $\dfrac{5\pi}{12} \times \dfrac{180}{\pi} = 75°$

     (b) $\dfrac{\pi}{24} \times \dfrac{180}{\pi} = 7.5°$

     (c) $3.4 \times \dfrac{180}{\pi} = 195°$

!   When working in radians with trigonometric functions on your calculator, ensure it is set in 'RAD' or 'R' mode.

Exercise

① Express the following angles in radians, leaving your answers in terms of π or to 3 significant figures as appropriate.

(i) $60°$ (ii) $45°$ (iii) $150°$ (iv) $200°$

(v) $44.4°$ (vi) $405°$ (vii) $270°$ (viii) $99°$

(ix) $300°$ (x) $720°$ (xi) $15°$ (xii) $3°$

② Express the following angles in degrees, rounding to 3 significant figures where appropriate.

(i) $\dfrac{\pi}{9}$ (ii) $\dfrac{2\pi}{15}$ (iii) 4^c (iv) $\dfrac{5\pi}{3}$

(v) $\dfrac{\pi}{7}$ (vi) $\dfrac{\pi}{20}$ (vii) 1.8^c (viii) $\dfrac{11\pi}{9}$

(ix) $\dfrac{7\pi}{2}$ (x) 5π (xi) $\dfrac{9\pi}{4}$ (xii) $\dfrac{17\pi}{12}$

In Chapters 1 and 6 of this book you use the trigonometric identities known as the **addition formulae** or **compound angle formulae**. The proofs of these identities are given in the A level Mathematics textbook.

These identities are:

$$\sin(\theta + \phi) \equiv \sin\theta \cos\phi + \cos\theta \sin\phi$$

$$\sin(\theta - \phi) \equiv \sin\theta \cos\phi - \cos\theta \sin\phi$$

$$\cos(\theta + \phi) \equiv \cos\theta \cos\phi - \sin\theta \sin\phi$$

$$\cos(\theta - \phi) \equiv \cos\theta \cos\phi + \sin\theta \sin\phi$$

Note the change of sign in the formulae for the cosine of the sum or difference of two angles:

$$\cos(\theta + \phi) \equiv \cos\theta \cos\phi - \sin\theta \sin\phi$$

$$\cos(\theta - \phi) \equiv \cos\theta \cos\phi + \sin\theta \sin\phi$$

Although these results are often referred to as 'formulae', they are in fact identities (as indicated by the identity symbol $\equiv$) and they are true for all values of θ and ϕ. However, it is common for the identity symbol to be replaced by an equals sign when the formulae are being used.

These identities will be used:

- in Chapter 1 to look at combinations of two rotations
- in Chapter 6 to look at multiplying two complex numbers in modulus-argument form.

Example

Use the compound angle formulae to find exact values for:

(i) $\sin 15°$

(ii) $\cos 75°$

Solution

(i) $\sin 15° = \sin(45° - 30°) = \sin 45° \cos 30° - \cos 45° \sin 30°$

$$= \frac{1}{\sqrt{2}} \times \frac{\sqrt{3}}{2} - \frac{1}{\sqrt{2}} \times \frac{1}{2}$$

$$= \frac{\sqrt{3}}{2\sqrt{2}} - \frac{1}{2\sqrt{2}}$$

$$= \frac{\sqrt{3}-1}{2\sqrt{2}} \text{ or } \frac{\sqrt{6}-\sqrt{2}}{4}$$

(ii) $\cos 75° = \cos(45° + 30°) = \cos 45° \cos 30° - \sin 45° \sin 30°$

$$= \frac{1}{\sqrt{2}} \times \frac{\sqrt{3}}{2} - \frac{1}{\sqrt{2}} \times \frac{1}{2}$$

This is the same as part (i) and so $\cos 75° = \dfrac{\sqrt{6}-\sqrt{2}}{4}$.

The exercise below is designed to familiarise you with these identities.

Exercise

① Use the compound angle formulae to write the following in surd form:

   (i) $\cos 15° = \cos(45° - 30°)$

   (ii) $\sin 105° = \sin(60° + 45°)$

   (iii) $\cos 105° = \cos(60° + 45°)$

   (iv) $\sin 165° = \sin(120° + 45°)$

② Simplify each of the following expressions, giving answers in surd form where possible:

   (i) $\sin 60° \cos 30° - \cos 60° \sin 30°$

   (ii) $\sin 40° \cos 50° + \cos 40° \sin 50°$

   (iii) $\cos 3\theta \cos \theta - \sin 3\theta \sin \theta$

   (iv) $\cos\left(\dfrac{\pi}{3}\right)\cos\left(\dfrac{\pi}{6}\right) + \sin\left(\dfrac{\pi}{3}\right)\sin\left(\dfrac{\pi}{6}\right)$

   (v) $2\sin\left(\dfrac{\pi}{4}\right)\cos\left(\dfrac{\pi}{6}\right) - 2\cos\left(\dfrac{\pi}{4}\right)\sin\left(\dfrac{\pi}{6}\right)$

   (vi) $\cos 47° \cos 13° - \sin 13° \sin 47°$

③ Expand and simplify the following expressions:

   (i) $\sin(\theta + 45°)$

   (ii) $\cos(2\theta - 30°)$

   (iii) $\sin\left(\theta - \dfrac{\pi}{6}\right)$

   (iv) $\cos\left(3\theta + \dfrac{\pi}{3}\right)$

Chapter 1

Discussion point (Page 1)

3, 2, 1, 0

Discussion point (Page 4)

When subtracting numbers, the order in which the numbers appear is important – changing the order changes the answer, for example: $3 - 6 \neq 6 - 3$. So subtraction of numbers is not commutative.

The grouping of the numbers is also important, for example $(13 - 5) - 2 \neq 13 - (5 - 2)$. Therefore subtraction of numbers is not associative.

Matrices follow the same rules for commutativity and associativity under addition and subtraction as numbers. Matrix addition is both commutative and associative, but matrix subtraction is not commutative or associative. This is true because addition and subtraction of each of the individual elements will determine whether the matrices are commutative or associative overall.

You can use more formal methods to prove these properties. For example, to show that matrix addition is commutative:

$$\begin{pmatrix} a & b \\ c & d \end{pmatrix} + \begin{pmatrix} e & f \\ g & h \end{pmatrix} = \begin{pmatrix} a+e & b+f \\ c+g & d+h \end{pmatrix} = \begin{pmatrix} e+a & f+b \\ g+c & h+d \end{pmatrix} = \begin{pmatrix} e & f \\ g & h \end{pmatrix} + \begin{pmatrix} a & b \\ c & d \end{pmatrix}$$

Addition of numbers is commutative

Exercise 1.1 (Page 4)

1 (i) 3×2 (ii) 3×3 (iii) 1×2
   (iv) 5×1 (v) 2×4 (vi) 3×2

2 (i) $\begin{pmatrix} 5 & -8 \\ 2 & -3 \end{pmatrix}$ (ii) $\begin{pmatrix} 3 & 1 & -4 \\ 4 & 2 & 12 \end{pmatrix}$

   (iii) $\begin{pmatrix} -8 & 5 \\ -3 & 7 \end{pmatrix}$

   (iv) Non-conformable (v) $\begin{pmatrix} -3 & -9 & 14 \\ 0 & 0 & 4 \end{pmatrix}$

   (vi) $\begin{pmatrix} 4 \\ 12 \\ 20 \end{pmatrix}$ (vii) $\begin{pmatrix} 9 & 7 & -17 \\ 10 & 5 & 28 \end{pmatrix}$

   (viii) Non-conformable

   (ix) $\begin{pmatrix} -15 & 8 \\ -4 & 3 \end{pmatrix}$

3 (i) $\begin{pmatrix} 0 & 2 & 1 & 0 \\ 1 & 0 & 2 & 1 \\ 0 & 2 & 0 & 2 \\ 1 & 0 & 1 & 0 \end{pmatrix}$ (ii) $\begin{pmatrix} 0 & 0 & 2 & 2 \\ 1 & 0 & 0 & 0 \\ 2 & 0 & 0 & 1 \\ 0 & 0 & 2 & 0 \end{pmatrix}$

(iii)

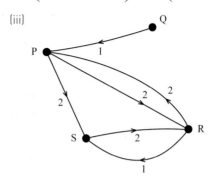

4 $w = 2, x = -6, y = -2, z = 2$

5 $p = -1$ or 6, $q = \pm\sqrt{5}$

6 (i) $\begin{pmatrix} 1 & 0 & 1 & 4 & 4 \\ 0 & 0 & 1 & 0 & 2 \\ 1 & 1 & 0 & 7 & 5 \\ 0 & 1 & 0 & 3 & 3 \end{pmatrix}$

   $\begin{pmatrix} 3 & 1 & 1 & 10 & 7 \\ 0 & 0 & 4 & 2 & 10 \\ 3 & 1 & 1 & 11 & 8 \\ 1 & 2 & 1 & 8 & 6 \end{pmatrix}$

   (ii) $\begin{pmatrix} 1 & 0 & 0 & 2 & 1 \\ 1 & 1 & 0 & 3 & 2 \\ 0 & 0 & 1 & 1 & 2 \\ 0 & 1 & 1 & 2 & 3 \end{pmatrix}$

   City 2 vs United 1
   Rangers 2 vs Town 1
   Rangers 1 vs /United 1

7 (i) $\begin{pmatrix} 15 & 3 & 7 & 15 \\ 5 & 9 & 15 & -3 \\ 19 & 10 & 9 & 3 \end{pmatrix}$

The matrix represents the number of jackets left in stock after all the orders have been dispatched. The negative element indicates there was not enough of that type of jacket in stock to fulfil the order.

(ii) $\begin{pmatrix} 20 & 13 & 17 & 20 \\ 15 & 19 & 20 & 12 \\ 19 & 10 & 14 & 8 \end{pmatrix}$

(iii) $\begin{pmatrix} 12 & 30 & 18 & 0 \\ 6 & 18 & 24 & 36 \\ 30 & 0 & 12 & 18 \end{pmatrix}$

Probably not very realistic, as a week is quite a short time.

Discussion point (Page 8)

The dimensions of the matrices are $\mathbf{A}$ (3×3), $\mathbf{B}$ (3×2) and $\mathbf{C}$ (2×2). The conformable products are $\mathbf{AB}$ and $\mathbf{BC}$. Both of these products would have dimension (3×2), even though the original matrices are not the same sizes.

Activity 1.1 (Page 9)

$\mathbf{AB} = \begin{pmatrix} 2 & -1 \\ 3 & 4 \end{pmatrix}\begin{pmatrix} -4 & 0 \\ -2 & 1 \end{pmatrix} = \begin{pmatrix} -6 & -1 \\ -20 & 4 \end{pmatrix}$

$\mathbf{BA} = \begin{pmatrix} -4 & 0 \\ -2 & 1 \end{pmatrix}\begin{pmatrix} 2 & -1 \\ 3 & 4 \end{pmatrix} = \begin{pmatrix} -8 & 4 \\ -1 & 6 \end{pmatrix}$

These two matrices are not equal and so matrix multiplication is not usually commutative. There are some exceptions, for example if

$\mathbf{C} = \begin{pmatrix} 2 & 0 \\ 0 & 2 \end{pmatrix}$ and $\mathbf{D} = \begin{pmatrix} 3 & 3 \\ -1 & -1 \end{pmatrix}$ then

$\mathbf{CD} = \mathbf{DC} = \begin{pmatrix} 6 & 6 \\ -2 & -2 \end{pmatrix}$.

Activity 1.2 (Page 10)

(i) $\mathbf{AB} = \begin{pmatrix} -6 & -1 \\ -20 & 4 \end{pmatrix}$

(ii) $\mathbf{BC} = \begin{pmatrix} -4 & -8 \\ 0 & -1 \end{pmatrix}$

(iii) $(\mathbf{AB})\mathbf{C} = \begin{pmatrix} -8 & -15 \\ -12 & -28 \end{pmatrix}$

(iv) $\mathbf{A}(\mathbf{BC}) = \begin{pmatrix} -8 & -15 \\ -12 & -28 \end{pmatrix}$

$(\mathbf{AB})\mathbf{C} = \mathbf{A}(\mathbf{BC})$ so matrix multiplication is associative in this case

To produce a general proof, use general matrices such as

$\mathbf{A} = \begin{pmatrix} a & b \\ c & d \end{pmatrix}$, $\mathbf{B} = \begin{pmatrix} e & f \\ g & h \end{pmatrix}$ and

$\mathbf{C} = \begin{pmatrix} i & j \\ k & l \end{pmatrix}$.

$\mathbf{AB} = \begin{pmatrix} a & b \\ c & d \end{pmatrix}\begin{pmatrix} e & f \\ g & h \end{pmatrix} = \begin{pmatrix} ae + bg & af + bh \\ ce + dg & cf + dh \end{pmatrix}$,

$\mathbf{BC} = \begin{pmatrix} e & f \\ g & h \end{pmatrix}\begin{pmatrix} i & j \\ k & l \end{pmatrix} = \begin{pmatrix} ei + fk & ej + fl \\ gi + hk & gj + hl \end{pmatrix}$

and so

$(\mathbf{AB})\mathbf{C} = \begin{pmatrix} ae + bg & af + bh \\ ce + dg & cf + dh \end{pmatrix}\begin{pmatrix} i & j \\ k & l \end{pmatrix}$

$\begin{pmatrix} aei + bgi + afk + bhk & aej + bgj + afl + bhl \\ cei + cfk + dgi + dhk & cej + cfl + dgj + dhl \end{pmatrix}$

and

$\mathbf{A}(\mathbf{BC}) = \begin{pmatrix} a & b \\ c & d \end{pmatrix}\begin{pmatrix} ei + fk & ej + fl \\ gi + hk & gj + hl \end{pmatrix}$

$= \begin{pmatrix} aei + afk + bgi + bhk & aej + afl + bgj + bhl \\ cei + dgi + cfk + dhk & cej + dgj + cfl + dhl \end{pmatrix}$

Since $(\mathbf{AB})\mathbf{C} = \mathbf{A}(\mathbf{BC})$ matrix multiplication is associative and the product can be written without brackets as $\mathbf{ABC}$.

Exercise 1.2 (Page 10)

1 (i) (a) 3×3 (b) 1×3 (c) 2×3 (d) 2×4
      (e) 2×1 (f) 3×5
   (ii) (a) non-conformable
        (b) 3×5
        (c) non-conformable
        (d) 2×3
        (e) non-conformable

2 (i) $\begin{pmatrix} 21 & 6 \\ 31 & 13 \end{pmatrix}$ (ii) $\begin{pmatrix} -30 & -15 \end{pmatrix}$

(iii) $\begin{pmatrix} -54 \\ -1 \end{pmatrix}$

3 $\mathbf{AB} = \begin{pmatrix} 3 & -56 \\ 20 & -73 \end{pmatrix}$, $\mathbf{BA} = \begin{pmatrix} -25 & 8 \\ 28 & -45 \end{pmatrix}$

$\mathbf{AB} \neq \mathbf{BA}$ so matrix multiplication is non-commutative.

4 (i) $\begin{pmatrix} -7 & 26 \\ 2 & 34 \end{pmatrix}$ (ii) $\begin{pmatrix} 5 & 25 \\ 16 & 22 \end{pmatrix}$

(iii) $\begin{pmatrix} 31 & 0 \\ 65 & 18 \end{pmatrix}$ (iv) $\begin{pmatrix} 26 & 37 & 16 \\ 14 & 21 & 28 \\ -8 & -11 & 2 \end{pmatrix}$

(v) non-conformable (vi) $\begin{pmatrix} 28 & -18 \\ 26 & 2 \\ 16 & 25 \end{pmatrix}$

5 $\begin{pmatrix} -38 & -136 & -135 \\ 133 & 133 & 100 \\ 273 & 404 & 369 \end{pmatrix}$

6 (i) $\begin{pmatrix} 2x^2 + 12 & -9 \\ -4 & 3 \end{pmatrix}$ (ii) $x = 2$ or 3

(iii) $\mathbf{BA} = \begin{pmatrix} 8 & 12 \\ 8 & 15 \end{pmatrix}$ or $\begin{pmatrix} 18 & 18 \\ 12 & 15 \end{pmatrix}$

7 (i) (a) $\begin{pmatrix} 4 & 3 \\ 0 & 1 \end{pmatrix}$ (b) $\begin{pmatrix} 8 & 7 \\ 0 & 1 \end{pmatrix}$

(c) $\begin{pmatrix} 16 & 15 \\ 0 & 1 \end{pmatrix}$ (ii) $\begin{pmatrix} 2^n & 2^n - 1 \\ 0 & 1 \end{pmatrix}$

8 (i) $\begin{pmatrix} 1 & 1 & 2 & 0 \\ 1 & 0 & 1 & 0 \\ 1 & 1 & 0 & 2 \\ 0 & 0 & 1 & 0 \end{pmatrix}$

(ii) $\begin{pmatrix} 4 & 3 & 3 & 4 \\ 2 & 2 & 2 & 2 \\ 2 & 1 & 5 & 0 \\ 1 & 1 & 0 & 2 \end{pmatrix}$ $\mathbf{M}^2$ represents the number of two-stage routes between each pair of resorts.

(iii) $\mathbf{M}^3$ would represent the number of three-stage routes between each pair of resorts.

9 (i) $\begin{pmatrix} 8 + 4x & -20 + x^2 \\ x - 8 & -3 - 3x \end{pmatrix}$

(ii) $x = -3$ or 4

(iii) $\begin{pmatrix} -4 & -11 \\ -11 & 6 \end{pmatrix}$ or $\begin{pmatrix} 24 & -4 \\ -4 & -15 \end{pmatrix}$

10 (i) $\mathbf{D} = \begin{pmatrix} 1 & 1 & 1 & 1 \end{pmatrix}$

$\mathbf{DA} = \begin{pmatrix} 299 & 199 & 270 & 175 & 114 \end{pmatrix}$

(ii) $\mathbf{F} = \begin{pmatrix} 1 \\ 1 \\ 1 \\ 1 \\ 1 \end{pmatrix}$, $\mathbf{AF} = \begin{pmatrix} 229 \\ 231 \\ 263 \\ 334 \end{pmatrix}$

(iii) $\mathbf{S} = \begin{pmatrix} 1 \\ 0 \\ 0 \\ 0 \\ 1 \end{pmatrix}$, $\mathbf{DAS} = (413)$,

$\mathbf{C} = \begin{pmatrix} 0 \\ 1 \\ 1 \\ 1 \\ 0 \end{pmatrix}$, $\mathbf{DAC} = (644)$

(iv) $\mathbf{P} = \begin{pmatrix} 0.95 \\ 0.95 \\ 1.05 \\ 1.15 \\ 1.15 \end{pmatrix}$,

$\mathbf{DAP} = (1088.95) = \pounds1088.95$

11 (i) $\begin{pmatrix} b \\ a \\ c \end{pmatrix}$ (ii) $\begin{pmatrix} 1 & 0 & 0 \\ 0 & 0 & 1 \\ 0 & 1 & 0 \end{pmatrix}$

(iii) $\begin{pmatrix} 0 & 1 & 0 \\ 0 & 0 & 1 \\ 1 & 0 & 0 \end{pmatrix}$, $\begin{pmatrix} b \\ c \\ a \end{pmatrix}$

(iv) $\begin{pmatrix} 0 & 0 & 1 \\ 1 & 0 & 0 \\ 0 & 1 & 0 \end{pmatrix}$, $\begin{pmatrix} c \\ a \\ b \end{pmatrix}$

(v) $\begin{pmatrix} 1 & 0 & 0 \\ 0 & 1 & 0 \\ 0 & 0 & 1 \end{pmatrix}$ The strands are back in the original order at the end of Stage 6.

Discussion point (Page 17)

The image of the unit vector $\begin{pmatrix} 1 \\ 0 \end{pmatrix}$ is $\begin{pmatrix} a \\ c \end{pmatrix}$ and the image of the unit vector $\begin{pmatrix} 0 \\ 1 \end{pmatrix}$ is $\begin{pmatrix} b \\ d \end{pmatrix}$.

Activity 1.3 (Page 17)

The diagram below shows the unit square with two of its sides along the unit vectors **i** and **j**. It is rotated by 45° about the origin.

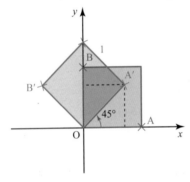

You can use trigonometry to find the images of the unit vectors **i** and **j**.

For A′, the x-coordinate satisfies $\cos 45 = \frac{x}{1}$ so $x = \cos 45 = \frac{1}{\sqrt{2}}$.

In a similar way, the y-coordinate of A′ is $\frac{1}{\sqrt{2}}$.

For B′, the symmetry of the diagram shows that the x-coordinate is $-\frac{1}{\sqrt{2}}$ and the y-coordinate is $\frac{1}{\sqrt{2}}$.

Hence, the image of $\begin{pmatrix} 1 \\ 0 \end{pmatrix}$ is $\begin{pmatrix} \frac{1}{\sqrt{2}} \\ \frac{1}{\sqrt{2}} \end{pmatrix}$ and the image of $\begin{pmatrix} 0 \\ 1 \end{pmatrix}$ is

$\begin{pmatrix} -\frac{1}{\sqrt{2}} \\ \frac{1}{\sqrt{2}} \end{pmatrix}$ and so the matrix representing an anticlockwise rotation of 45° about the origin is $\begin{pmatrix} \frac{1}{\sqrt{2}} & -\frac{1}{\sqrt{2}} \\ \frac{1}{\sqrt{2}} & \frac{1}{\sqrt{2}} \end{pmatrix}$.

Rotations of 45° clockwise about the origin and 135° anticlockwise about the origin are also represented by matrices involving $\pm\frac{1}{\sqrt{2}}$.

This is due to the symmetry about the origin.

(i) The diagram for a 45° clockwise rotation about the origin is shown below.

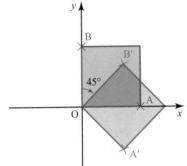

The image of $\begin{pmatrix} 1 \\ 0 \end{pmatrix}$ is $\begin{pmatrix} \frac{1}{\sqrt{2}} \\ -\frac{1}{\sqrt{2}} \end{pmatrix}$ and the image of $\begin{pmatrix} 0 \\ 1 \end{pmatrix}$ is $\begin{pmatrix} \frac{1}{\sqrt{2}} \\ \frac{1}{\sqrt{2}} \end{pmatrix}$ and so the matrix representing an anticlockwise rotation of 45° about the origin is $\begin{pmatrix} \frac{1}{\sqrt{2}} & \frac{1}{\sqrt{2}} \\ -\frac{1}{\sqrt{2}} & \frac{1}{\sqrt{2}} \end{pmatrix}$.

(ii) The diagram for a 135° anticlockwise rotation about the origin is shown below.

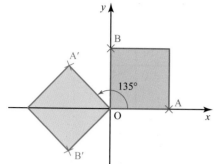

The image of $\begin{pmatrix} 1 \\ 0 \end{pmatrix}$ is $\begin{pmatrix} -\dfrac{1}{\sqrt{2}} \\ \dfrac{1}{\sqrt{2}} \end{pmatrix}$ and the image of

$\begin{pmatrix} 0 \\ 1 \end{pmatrix}$ is $\begin{pmatrix} -\dfrac{1}{\sqrt{2}} \\ -\dfrac{1}{\sqrt{2}} \end{pmatrix}$ and so the matrix representing

an anticlockwise rotation of 45° about the origin is

$\begin{pmatrix} -\dfrac{1}{\sqrt{2}} & -\dfrac{1}{\sqrt{2}} \\ \dfrac{1}{\sqrt{2}} & -\dfrac{1}{\sqrt{2}} \end{pmatrix}.$

Discussion point (Page 18)

The matrix for a rotation of $\theta°$ clockwise about the

origin is $\begin{pmatrix} \cos\theta & \sin\theta \\ -\sin\theta & \cos\theta \end{pmatrix}$

Activity 1.4 (Page 19)

(i) The diagram below shows the effect of the matrix

$\begin{pmatrix} 2 & 0 \\ 0 & 1 \end{pmatrix}$ on the unit vectors **i** and **j**.

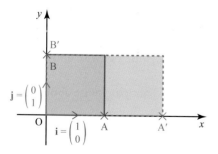

You can see that the vector **i** has image $\begin{pmatrix} 2 \\ 0 \end{pmatrix}$ and the vector **j** is unchanged. Therefore this matrix represents a stretch of scale factor 2 parallel to the x-axis.

(ii) The diagram below shows the effect of the matrix

$\begin{pmatrix} 1 & 0 \\ 0 & 5 \end{pmatrix}$ on the unit vectors **i** and **j**.

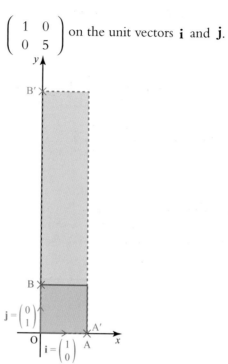

You can see that the vector **i** is unchanged and the vector **j** has image $\begin{pmatrix} 0 \\ 5 \end{pmatrix}$. Therefore this matrix represents a stretch of scale factor 5 parallel to the y-axis.

The matrix $\begin{pmatrix} m & 0 \\ 0 & 1 \end{pmatrix}$ represents a stretch of scale factor m

parallel to the x-axis.

The matrix $\begin{pmatrix} 3 & 0 \\ 0 & 3 \end{pmatrix}$ represents a stretch of scale factor n

parallel to the y-axis.

Exercise 1.3 (Page 21)

1 (i) (a)

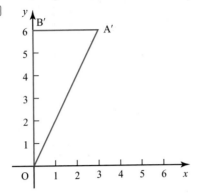

(b) A' = (3, 6), B' = (0, 6)

(c) $x' = 3x$, $y' = 3y$

(d) $\begin{pmatrix} 3 & 0 \\ 0 & 3 \end{pmatrix}$

(ii) (a)

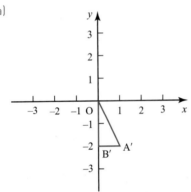

(b) A' = (1, −2), B' = (0, −2)

(c) $x' = x$, $y' = -y$

(d) $\begin{pmatrix} 1 & 0 \\ 0 & -1 \end{pmatrix}$

(iii) (a)

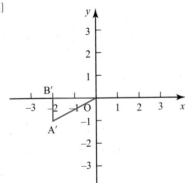

(b) A' = (−2, −1), B' = (−2, 0)

(c) $x' - y$ $y' = -x$

(d) $\begin{pmatrix} 0 & -1 \\ -1 & 0 \end{pmatrix}$

(iv) (a)

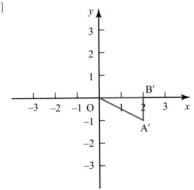

(b) A' = (2, −1), B' = (2, 0)

(c) $x' = y$, $y' = -x$

(d) $\begin{pmatrix} 0 & 1 \\ -1 & 0 \end{pmatrix}$

(v) (a)

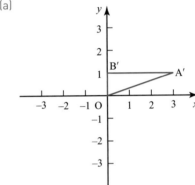

(b) A' = (3, 1), B' = (0, 1)

(c) $x' = 3x$, $y' = \frac{1}{2}y$

(d) $\begin{pmatrix} 3 & 0 \\ 0 & \frac{1}{2} \end{pmatrix}$

2 (i) Reflection in the x-axis

(ii) Reflection in the line $y = -x$

(iii) Stretch of factor 2 parallel to the x-axis and stretch factor 3 parallel to the y-axis

(iv) Enlargement, scale factor 4, centre the origin

(v) Rotation of 90° clockwise (or 270° anticlockwise) about the origin

3 (i) Rotation of 60° anticlockwise about the origin

(ii) Rotation of 55° anticlockwise about the origin

(iii) Rotation of 135° clockwise about the origin

(iv) Rotation of 150° anticlockwise about the origin

4 (i) $\begin{pmatrix} 0 & -1 & 0 \\ 1 & 0 & 0 \\ 0 & 0 & 1 \end{pmatrix}$ (ii) $\begin{pmatrix} 1 & 0 & 0 \\ 0 & -1 & 0 \\ 0 & 0 & 1 \end{pmatrix}$

(iii) $\begin{pmatrix} 1 & 0 & 0 \\ 0 & -1 & 0 \\ 0 & 0 & -1 \end{pmatrix}$ (iv) $\begin{pmatrix} 0 & 0 & -1 \\ 0 & 1 & 0 \\ 1 & 0 & 0 \end{pmatrix}$

5 $A'(4, 5)$, $B'(7, 9)$, $C'(3, 4)$. The original square and the image both have an area of one square unit.

6 (i)

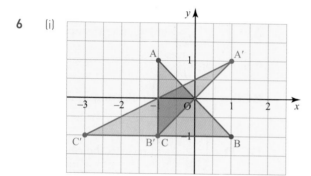

(ii) The gradient of $A'C'$ is $\frac{1}{2}$, which is the reciprocal of the top right-hand entry of the matrix **M**.

7 (i) Rotation of 90° clockwise about the x-axis
(ii) Enlargement scale factor 3, centre $(0, 0)$
(iii) Reflection in the plane $z = 0$
(iv) Three-way stretch of factor 2 in the x-direction, factor 3 in the y-direction and factor 0.5 in the z-direction

8 (i) $\begin{pmatrix} 1 & 0 & 0 \\ 0 & 1 & 0 \\ 0 & 0 & -1 \end{pmatrix}$ (ii) $\begin{pmatrix} -1 & 0 & 0 \\ 0 & 1 & 0 \\ 0 & 0 & -1 \end{pmatrix}$

9 $(x, y) \rightarrow (x, x)$

The matrix for the transformation is $\begin{pmatrix} 1 & 0 \\ 1 & 0 \end{pmatrix}$.

10 (i) Any matrix of the form $\begin{pmatrix} 5 & 0 \\ 0 & k \end{pmatrix}$ or

$\begin{pmatrix} k & 0 \\ 0 & 5 \end{pmatrix}$.

If $k = 5$ the rectangle would be a square.

(ii) $\begin{pmatrix} \sqrt{2} & 1 \\ 0 & 1 \end{pmatrix}$, $\begin{pmatrix} 1 & 0 \\ 1 & \sqrt{2} \end{pmatrix}$,

$\begin{pmatrix} 1 & \sqrt{2} \\ 1 & 0 \end{pmatrix}$ or $\begin{pmatrix} 0 & 1 \\ \sqrt{2} & 1 \end{pmatrix}$

(iii) $\begin{pmatrix} 7 & \frac{3\sqrt{3}}{2} \\ 0 & \frac{3}{2} \end{pmatrix}$, $\begin{pmatrix} 0 & \frac{3}{2} \\ 7 & \frac{3\sqrt{3}}{2} \end{pmatrix}$,

$\begin{pmatrix} \frac{3\sqrt{3}}{2} & 7 \\ \frac{3}{2} & 0 \end{pmatrix}$ or $\begin{pmatrix} 0 & \frac{3}{2} \\ 7 & \frac{3\sqrt{3}}{2} \end{pmatrix}$

Discussion point (Page 24)

(i) **BA** represents a reflection in the line $y = x$

(ii) The transformation A is represented by the matrix $\mathbf{A} = \begin{pmatrix} 1 & 0 \\ 0 & -1 \end{pmatrix}$ and the transformation B is represented by the matrix

$\mathbf{B} = \begin{pmatrix} 0 & -1 \\ 1 & 0 \end{pmatrix}$. The matrix product

$\mathbf{BA} = \begin{pmatrix} 0 & -1 \\ 1 & 0 \end{pmatrix}\begin{pmatrix} 1 & 0 \\ 0 & -1 \end{pmatrix} = \begin{pmatrix} 0 & 1 \\ 1 & 0 \end{pmatrix}$.

This is the matrix which represents a reflection in the line $y = x$.

Activity 1.5 (Page 24)

(i) $\mathbf{P'} = \begin{pmatrix} a & b \\ c & d \end{pmatrix}\begin{pmatrix} x \\ y \end{pmatrix} = \begin{pmatrix} ax + by \\ cx + dy \end{pmatrix}$

(ii) $\mathbf{P''} = \begin{pmatrix} p & q \\ r & s \end{pmatrix}\begin{pmatrix} ax + by \\ cx + dy \end{pmatrix}$

$= \begin{pmatrix} pax + pby + qcx + qdy \\ rax + rby + scx + sdy \end{pmatrix}$

(iii)

$$\mathbf{U} = \begin{pmatrix} p & q \\ r & s \end{pmatrix}\begin{pmatrix} a & b \\ c & d \end{pmatrix} = \begin{pmatrix} pa + qc & pb + qd \\ ra + sc & rb + sd \end{pmatrix}$$

and so

$$\mathbf{UP} = \begin{pmatrix} pa + qc & pb + qd \\ ra + sc & rb + sd \end{pmatrix}\begin{pmatrix} x \\ y \end{pmatrix}$$

$$= \begin{pmatrix} pax + qcx + pby + rdy \\ rax + scx + rby + sdy \end{pmatrix}. \text{ Therefore } \mathbf{UP} = \mathbf{P}''$$

Discussion point (Page 24)

AB represents 'carry out transformation B followed by transformation A.

(AB)C represents 'carry out transformation C followed by transformation AB, i.e. 'carry out C followed by B followed by A'.

BC represents 'carry out transformation C followed by transformation B'.

A(BC) represents 'carry out transformation BC followed by transformation A, i.e. carry out C followed by B followed by A'.

Activity 1.6 (Page 25)

(i) $\mathbf{A} = \begin{pmatrix} \cos\theta & -\sin\theta \\ \sin\theta & \cos\theta \end{pmatrix}$,

$\mathbf{B} = \begin{pmatrix} \cos\phi & -\sin\phi \\ \sin\phi & \cos\phi \end{pmatrix}$

(ii)

$$\mathbf{BA} = \begin{pmatrix} \cos\theta\cos\phi - \sin\theta\sin\phi & -\sin\theta\cos\phi - \cos\theta\sin\phi \\ \sin\theta\cos\phi + \cos\theta\sin\phi & -\sin\theta\sin\phi + \cos\theta\cos\phi \end{pmatrix}$$

(iii) $\mathbf{C} = \begin{pmatrix} \cos(\theta + \phi) & -\sin(\theta + \phi) \\ \sin(\theta + \phi) & \cos(\theta + \phi) \end{pmatrix}$

(iv) $\sin(\theta + \phi) = \sin\theta\cos\phi + \cos\theta\sin\phi$

$\cos(\theta + \phi) = \cos\theta\cos\phi - \sin\theta\sin\phi$

(v) A rotation through angle θ followed by rotation through angle ϕ has the same effect as a rotation through angle ϕ followed by angle θ.

Exercise 1.4 (Page 26)

1 (i) **A**: enlargement centre (0,0), scale factor 3
      B: rotation 90° anticlockwise about (0,0)
      C: reflection in the x-axis
      D: reflection in the line $y = x$

(ii) $\mathbf{BC} = \begin{pmatrix} 0 & 1 \\ 1 & 0 \end{pmatrix}$, reflection in the line

$y = x$

$\mathbf{CB} = \begin{pmatrix} 0 & -1 \\ -1 & 0 \end{pmatrix}$, reflection in the line

$y = -x$

$\mathbf{DC} = \begin{pmatrix} 0 & -1 \\ 1 & 0 \end{pmatrix}$, rotation 90°

anticlockwise about $(0, 0)$

$\mathbf{A}^2 = \begin{pmatrix} 9 & 0 \\ 0 & 9 \end{pmatrix}$, enlargement centre $(0, 0)$,

scale factor 9

$\mathbf{BCB} = \begin{pmatrix} 1 & 0 \\ 0 & -1 \end{pmatrix}$, reflection in the x-axis

$\mathbf{DC}^2\mathbf{D} = \begin{pmatrix} 1 & 0 \\ 0 & 1 \end{pmatrix}$ returns the object to

its original position

(iii) For example, $\mathbf{B}^4$, $\mathbf{C}^2$ or $\mathbf{D}^2$

2 (i) $\mathbf{X} = \begin{pmatrix} 1 & 0 \\ 0 & -1 \end{pmatrix}$ $\mathbf{Y} = \begin{pmatrix} -1 & 0 \\ 0 & 1 \end{pmatrix}$

(ii) $\mathbf{XY} = \begin{pmatrix} -1 & 0 \\ 0 & -1 \end{pmatrix}$, rotation of 180°

about the origin

(iii) $\mathbf{YX} = \begin{pmatrix} -1 & 0 \\ 0 & -1 \end{pmatrix}$

(iv) When considering the effect on the unit vectors $\mathbf{i}$ and $\mathbf{j}$, as each transformation only affects one of the unit vectors the order of the transformations is not important in this case.

3 (i) $\mathbf{P} = \begin{pmatrix} -1 & 0 \\ 0 & -1 \end{pmatrix}$ $\mathbf{Q} = \begin{pmatrix} 0 & 1 \\ 1 & 0 \end{pmatrix}$

(ii) $\mathbf{PQ} = \begin{pmatrix} 0 & -1 \\ -1 & 0 \end{pmatrix}$, reflection in the line

$y = -x$

(iii) $\mathbf{QP} = \begin{pmatrix} 0 & -1 \\ -1 & 0 \end{pmatrix}$

(iv) The matrix **P** has the effect of making the coordinates of any point the negative of their original values,

i.e. $(x, y) \rightarrow (-x, -y)$

The matrix **Q** interchanges the coordinates,

i.e. $(x, y) \rightarrow (y, x)$

It does not matter what order these two transformations occur as the result will be the same

4 (i) $\mathbf{J} = \begin{pmatrix} 1 & 0 & 0 \\ 0 & 1 & 0 \\ 0 & 0 & -1 \end{pmatrix}$ $\mathbf{K} = \begin{pmatrix} 1 & 0 & 0 \\ 0 & 0 & -1 \\ 0 & 1 & 0 \end{pmatrix}$

$\mathbf{L} = \begin{pmatrix} -1 & 0 & 0 \\ 0 & 1 & 0 \\ 0 & 0 & 1 \end{pmatrix}$ $\mathbf{M} = \begin{pmatrix} 0 & 0 & 1 \\ 0 & 1 & 0 \\ -1 & 0 & 0 \end{pmatrix}$

(ii) (a) **LJ** (b) **MJ**
(c) **K²** (d) **JLK**

5 (i) $\begin{pmatrix} 8 & -4 \\ -3 & 12 \end{pmatrix}$ (ii) $(32, -33)$

6 Possible transformations are $\mathbf{B} = \begin{pmatrix} 0 & 1 \\ -1 & 0 \end{pmatrix}$,

which is a rotation of 90° clockwise about the origin, followed by

$\mathbf{A} = \begin{pmatrix} 3 & 0 \\ 0 & 1 \end{pmatrix}$, which is a stretch of scale factor

3 parallel to the x-axis. The order of these is important as performing **A** followed by **B** leads

to the matrix $\begin{pmatrix} 0 & 1 \\ -3 & 0 \end{pmatrix}$. Could also have

$\mathbf{B} = \begin{pmatrix} 1 & 0 \\ 0 & 3 \end{pmatrix}$, which represents a stretch of

factor 3 parallel to the y-axis, followed by

$\mathbf{A} = \begin{pmatrix} 0 & 1 \\ -1 & 0 \end{pmatrix}$, which represents a rotation of

90° clockwise about the origin; again the order is important.

7 $\mathbf{X} = \begin{pmatrix} \dfrac{1}{\sqrt{2}} & \dfrac{1}{\sqrt{2}} \\ \dfrac{1}{\sqrt{2}} & -\dfrac{1}{\sqrt{2}} \end{pmatrix}$

A matrix representing a rotation about the

origin has the form $\begin{pmatrix} \cos\theta & -\sin\theta \\ \sin\theta & \cos\theta \end{pmatrix}$ and so

the entries on the leading diagonal would be equal. That is not true for matrix **X** and so this cannot represent a rotation.

8 $\mathbf{Y} = \begin{pmatrix} 0 & 1 & 0 \\ 1 & 0 & 0 \\ 0 & 0 & 1 \end{pmatrix}$

9 (i) $\begin{pmatrix} 1 & 0 \\ 0 & 2 \end{pmatrix}$

(ii) A reflection in the x-axis and a stretch of scale factor 5 parallel to the x-axis

(iii) $\begin{pmatrix} 5 & 0 \\ 0 & -2 \end{pmatrix}$

Reflection in the x-axis; stretch of scale factor 5 parallel to the x-axis; stretch of scale factor 2 parallel to the y-axis. The outcome of these three transformations would be the same regardless of the order in which they are applied. There are six different possible orders.

(iv) $\begin{pmatrix} \dfrac{1}{5} & 0 \\ 0 & -\dfrac{1}{2} \end{pmatrix}$

10 (i) $\begin{pmatrix} 1 & -R_1 \\ 0 & 1 \end{pmatrix}$

(ii) $\begin{pmatrix} 1 & 0 \\ -\dfrac{1}{R_2} & 1 \end{pmatrix}$

(iii) $\begin{pmatrix} 1 & -R_1 \\ -\dfrac{1}{R_2} & \dfrac{R_1}{R_2} + 1 \end{pmatrix}$

(iv) $\begin{pmatrix} 1 + \dfrac{R_1}{R_2} & -R_1 \\ -\dfrac{1}{R_2} & 1 \end{pmatrix}$

The effect of Type B followed by Type A is different to that of Type A followed by Type B.

11    $a = \sqrt{\dfrac{\sqrt{2} + 2}{4}}$ and $b = \sqrt{\dfrac{1}{2(\sqrt{2} + 2)}}$

D represents an anticlockwise rotation of 22.5° about the origin.

By comparison to the matrix

$\begin{pmatrix} \cos\theta & -\sin\theta \\ \sin\theta & \cos\theta \end{pmatrix}$ for an anticlockwise

rotation of θ about the origin, a and b are the exact values of $\cos 22.5°$ and $\sin 22.5°$ respectively.

Discussion point (Page 28)

In a reflection, all points on the mirror line map to themselves.

In a rotation, only the centre of rotation maps to itself.

Exercise 1.5 (Page 31)

1   (i) Points of the form $(\lambda, -2\lambda)$

(ii) $(0, 0)$

(iii) Points of the form $(\lambda, -3\lambda)$

(iv) Points of the form $(2\lambda, 3\lambda)$

2   (i) x-axis, y-axis, lines of the form $y = mx$

(ii) x-axis, y-axis, lines of the form $y = mx$

(iii) no invariant lines

(iv) $y = x$, lines of the form $y = -x + c$

(v) $y = -x$, lines of the form $y = x + c$

3   (i) Any points on the line $y = \dfrac{1}{2}x$, for example $(0, 0)$, $(2, 1)$ and $(3, 1.5)$

(ii) $y = \dfrac{1}{2}x$

(iii) Any line of the form $y = -2x + c$

(iv) Using the method of Example 1.11 leads to the equations

$2m^2 + 3m - 2 = 0 \Rightarrow m = 0.5$ or -2

$(4 + 2m)c = 0 \Rightarrow m = -2$ or $c = 0$

If $m = 0.5$ then $c = 0$ so $y = \dfrac{1}{2}x$ is invariant.

If $m = -2$ then c can take any value and so $y = -2x + c$ is an invariant line.

4   (i) Solving $\begin{pmatrix} 4 & 11 \\ 11 & 4 \end{pmatrix} \begin{pmatrix} x \\ y \end{pmatrix} = \begin{pmatrix} x \\ y \end{pmatrix}$ leads

to the equations $y = -\dfrac{3x}{11}$ and $y = -\dfrac{11x}{3}$.

The only point that satisfies both of these is $(0, 0)$.

(ii) $y = x$ and $y = -x$

5   (i) $y = x$, $y = -\dfrac{9}{4}x$

(ii)

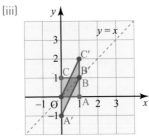

6   (i) $y = x$ (ii) $y = x$

(iii)

9   (i) $x' = x + a$, $y' = y + b$

(iii) (c) $a = -2b$

Chapter 2

Discussion point (Page 35)

$\mathbb{R}$ Real numbers – any number which is not complex

$\mathbb{Q}$ Rational numbers – numbers which can be expressed exactly as a fraction

$\mathbb{Z}$ Integers – positive or negative whole numbers, including zero

$\mathbb{N}$ Natural numbers – non-negative whole numbers (although there is some debate amongst mathematicians as to whether zero should be included!)

Discussion point (Page 36)

Any real number is either rational or irrational. This means that all real numbers will either lie inside the set of rational numbers, or inside the set of real numbers but outside the set of rational numbers. Therefore no separate set is needed for irrational numbers.

The symbol $\overline{\mathbb{Q}}$ is used for irrational numbers – numbers which cannot be expressed exactly as a fraction, such as π.

Activity 2.1 (Page 36)

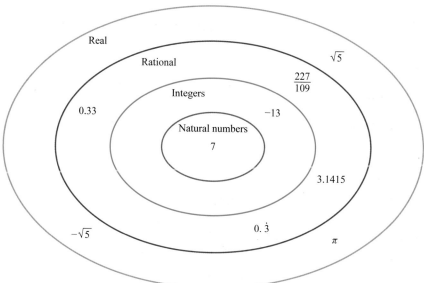

Activity 2.2 (Page 36)

(i) $x = 2$ Natural number (or integer)

(ii) $x = \dfrac{9}{7}$ Rational number

(iii) $x = \pm 3$ Integers

(iv) $x = -1$ Integer

(v) $x = 0,\ -7$ Integers

Discussion point (Page 37)

You know $i^2 = -1$

$i^3 = i^2 \times i = -1 \times i = -i$

$i^4 = i^2 \times i^2 = -1 \times -1 = 1$

$i^5 = i^4 \times i = 1 \times i = i$

$i^6 = i^5 \times i = i \times i = -1$

$i^7 = i^6 \times i = -1 \times i = -i$

The powers of i form a cycle:

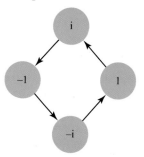

All numbers of the form i^{4n} are equal to 1.

All numbers of the form i^{4n+1} are equal to i.

All numbers of the form i^{4n+2} are equal to -1.

All numbers of the form i^{4n+3} are equal to $-i$.

Discussion point (Page 38)

$$\left(5 + \sqrt{-15}\right)\left(5 - \sqrt{-15}\right)$$
$$= 25 - 5\sqrt{-15} + 5\sqrt{-15} - (-15)$$
$$= 25 + 15$$
$$= 40$$

Discussion point (Page 38)

If the numerators and denominators of two fractions are equal then the fractions must also be equal.

However, it is possible for two fractions to be equal if the numerators and denominators are not equal, for example $\frac{3}{4} = \frac{6}{8}$.

Exercise 2.1 (Page 39)

1 (i) i   (ii) -1
   (iii) $-i$ (iv) 1

2 (i) $9 - i$ (ii) $-9 + 9i$
   (iii) $3 + 9i$ (iv) $-3 - i$

3 (i) $24 + 2i$ (ii) $-2 + 24i$
   (iii) $20 + 48i$ (iv) $38 - 18i$

4 (i) (a) 52   (b) 34   (c) 1768
   (ii) Multiplying a complex number by its conjugate gives a wholly real answer.

5 (i) $92 - 60i$ (ii) $-414 + 154i$

6 (i) $-1 \pm i$ (ii) $1 \pm 2i$ (iii) $2 \pm 3i$
   (iv) $-3 \pm 5i$ (v) $\frac{1}{2} \pm 2i$ (vi) $-2 \pm \sqrt{2}i$

7 $a = 1$ or $4, b = -1$ or 3
   The possible complex numbers are
   $1 + 9i$, $1 + i$, $16 + 9i$, $16 + i$

8 $a = 3$, $b = 5$ or $a = -3$, $b = -5$

9 $3 + 7i$ and $-3 - 7i$

10 (i)

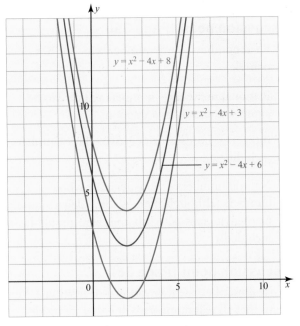

   (ii) (a) $x = 1$, $x = 3$
         (b) $x = 2 \pm \sqrt{2}i$
         (c) $2 \pm 2i$
   (iii) The roots all occur in pairs that are of the form $x = 2 \pm k$ for some value k where k is either a real number or a real multiple of i.

11 $a = -7$, $b = 11$
    You cannot assume the second root is the conjugate of $2 + i$ as the coefficients of the equation are not real.
    The second root is $5 - 2i$.

Activity 2.3 (Page 41)

$z + z^* = (x + yi) + (x - yi) = 2x$ which is real

$zz^* = (x + yi)(x - yi) = x^2 - xyi + yxi - y^2i^2 = x^2 + y^2$
which is real

Discussion point (Page 42)

$\dfrac{1}{i} = \dfrac{1}{i} \times \dfrac{i}{i} = \dfrac{i}{-1} = -i$

$\dfrac{1}{i^2} = \dfrac{1}{i^2} \times \dfrac{i^2}{i^2} = \dfrac{-1}{1} = -1$

$$\frac{1}{i^3} = \frac{1}{i^3} \times \frac{i^3}{i^3} = \frac{-i}{-1} = i$$

$$\frac{1}{i^4} = \frac{1}{i^4} \times \frac{i^4}{i^4} = \frac{1}{1} = 1$$

All numbers of the form $\frac{1}{i^{4n}}$ are equal to 1.

All numbers for the form $\frac{1}{i^{4n+1}}$ are equal to $-i$.

All numbers of the form $\frac{1}{i^{4n+2}}$ are equal to -1.

All numbers of the form $\frac{1}{i^{4n+3}}$ are equal to i.

Exercise 2.2 (Page 42)

1. (i) $\frac{21}{50} + \frac{3}{50}i$ (ii) $\frac{21}{50} - \frac{3}{50}i$

   (iii) $-\frac{3}{50} + \frac{21}{50}i$ (iv) $\frac{3}{50} + \frac{21}{50}i$

2. (i) $-\frac{9}{13} + \frac{19}{13}i$ (ii) $-\frac{9}{34} - \frac{19}{34}i$

   (iii) $-\frac{9}{13} - \frac{19}{13}i$ (iv) $-\frac{9}{34} + \frac{19}{34}i$

3. (i) $\frac{94}{25} + \frac{158}{25}i$ (ii) $\frac{204}{625} + \frac{253}{625}i$

4. (i) 6   (ii) 85

   (iii) 12   (iv) 45

   (v) -4 (vi) 45

5. (i) 2   (ii) 3

   (iii) $2 - 3i$ (iv) $6 + 4i$

   (v) $8 + i$ (vi) $-4 - 7i$

6. (i) 0   (ii) 0

   (iii) -39 (iv) $-46 + 9i$

   (v) $-46 - 9i$ (vi) $52i$

7. (i) $\frac{348}{61} + \frac{290}{61}i$ (ii) $\frac{322}{29} - \frac{65}{29}i$

   (iii) $-\frac{600}{3721} + \frac{110}{3721}i$

8. (i) $2 - i$ (ii) 1

   (iii) $3 + i$ (iv) $-\frac{35}{34} + \frac{149}{34}i$

9. $a = -\frac{23}{13}$ $b = -\frac{15}{13}$

10. $a = 9,\ b = 11$

11. (i) $\frac{10}{89}$ (ii) $\frac{10}{89}$

12. $\dfrac{2x}{x^2 + y^2}$

14. $a = 2,\ b = 2$

15. $z = 0,\quad z = 2,\quad z = -1 \pm \sqrt{3}i$

16. $z = 8 - 6i,\ w = 6 - 5i$

Discussion point (Page 43)

A complex number has a real component and an imaginary component. It is not possible to illustrate two components using a single number line.

Activity 2.4 (Page 44)

(i)

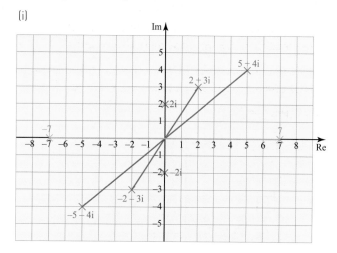

The points representing z and $-z$ have half turn rotational symmetry about the origin.

(ii)

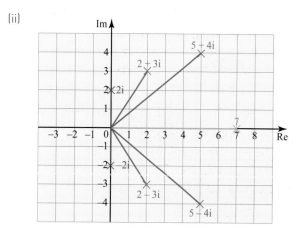

The points representing z and z^* are reflections of each other in the real axis.

Exercise 2.3 (Page 46)

1

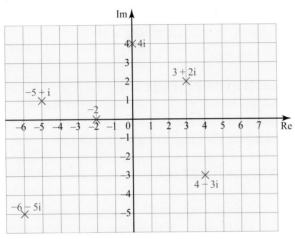

2

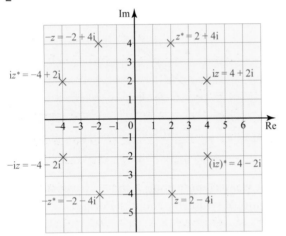

3

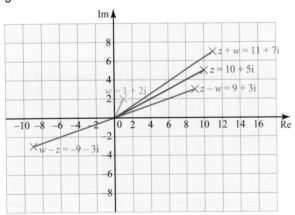

4 (i) $x^2 - 4x + 3 = 0$

(ii) $x^2 - 4x + 5 = 0$

(iii) $x^2 - 4x + 13 = 0$

(iv) All of the form $x^2 - 4x + k = 0$ where $k \in \mathbb{R}$

5

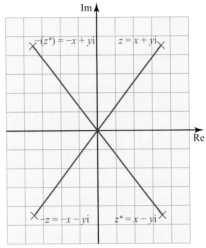

6 (i)

n	-1	0	1	2	3	4	5
z^n	$\dfrac{1}{2} - \dfrac{1}{2}i$	1	$1 + i$	$2i$	$-2 + 2i$	-4	$-4 - 4i$

(ii)

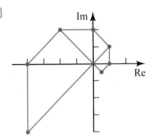

(iii)

n	-1	0	1	2	3	4	5
z^n	$\dfrac{1}{2} - \dfrac{1}{2}i$	1	$1 + i$	$2i$	$-2 + 2i$	-4	$-4 - 4i$
Distance from origin	$\dfrac{1}{\sqrt{2}}$	1	$\sqrt{2}$	2	$2\sqrt{2}$	4	$4\sqrt{2}$

(iv) The half squares formed are enlarged by a factor of $\sqrt{2}$ and rotated through $45°$ each time.

7 (i) $r = \sqrt{a^2 + b^2}$

$zz^* = (a + bi)(a - bi) = a^2 + b^2 = r^2$

(ii) $s = \sqrt{c^2 + d^2}$

(iii) $zw = (a + bi)(c + di) = (ac - bd) + (bc + ad)i$

Distance from origin of zw is

$$\sqrt{(ac - bd)^2 + (bc + ad)^2} = \sqrt{a^2c^2 + b^2d^2 + b^2c^2 + a^2d^2}$$

$$= \sqrt{(a^2 + b^2)(c^2 + d^2)}$$

$$= \sqrt{a^2 + b^2}\sqrt{c^2 + d^2} = rs$$

Chapter 3

Discussion point (Page 49)

$4x^3 + x^2 - 4x - 1 = 0$

Looking at the graph you may suspect that $x = 1$ is a root. Setting $x = 1$ verifies this. The factor theorem tells you that $(x - 1)$ must be a factor, so factorise the cubic $(x - 1)(4x^2 + 5x + 1) = 0$. Now factorise the remaining quadratic factor: $(x - 1)(4x + 1)(x + 1) = 0$, so the roots are $x = 1, -\frac{1}{4}, -1$.

$4x^3 + x^2 + 4x + 1 = 0$

This does not have such an obvious starting point, but the graph suggests only one real root.
Comparing with previous example, you may spot that $x = -\frac{1}{4}$ might work, so you can factorise giving $(4x + 1)(x^2 + 1) = 0$. From this you can see that the other roots must be complex. $x^2 = -1$, so the three roots are $x = -\frac{1}{4}, \pm i$.

Activity 3.1 (Page 50)

Equation	Two roots	Sum of roots	Product of roots
(i) $z^2 - 3z + 2 = 0$	1, 2	3	2
(ii) $z^2 + z - 6 = 0$	2, -3	-1	-6
(iii) $z^2 - 6z + 8 = 0$	2, 4	6	8
(iv) $z^2 - 3z - 10 = 0$	-2, 5	3	-10
(v) $2z^2 - 3z + 1 = 0$	$\frac{1}{2}$, 1	$\frac{3}{2}$	$\frac{1}{2}$
(vi) $z^2 - 4z + 5 = 0$	$2 \pm i$	4	5

Discussion point (Page 50)

If the equation is $ax^2 + bx + c = 0$, the sum appears to be $-\frac{b}{a}$ and the product appears to be $\frac{c}{a}$.

Discussion point (Page 51)

You get back to the original quadratic equation.

Exercise 3.1 (Page 53)

1. (i) $\alpha + \beta = -\frac{7}{2}$, $\alpha\beta = 3$

   (ii) $\alpha + \beta = \frac{1}{5}$, $\alpha\beta = -\frac{1}{5}$

   (iii) $\alpha + \beta = 0$, $\alpha\beta = \frac{2}{7}$

   (iv) $\alpha + \beta = -\frac{24}{5}$, $\alpha\beta = 0$

   (v) $\alpha + \beta = -11$, $\alpha\beta = -4$

   (vi) $\alpha + \beta = -\frac{8}{3}$, $\alpha\beta = -2$

2. (i) $z^2 - 10z + 21 = 0$

   (ii) $z^2 - 3z - 4 = 0$

   (iii) $2z^2 + 19z + 45 = 0$

   (iv) $z^2 - 5z = 0$

   (v) $z^2 - 6z + 9 = 0$

   (vi) $z^2 - 6z + 13 = 0$

3. (i) $2z^2 + 15z - 81 = 0$

   (ii) $2z^2 - 5z - 9 = 0$

   (iii) $2z^2 + 13z + 9 = 0$

   (iv) $z^2 - 7z - 12 = 0$

4. (i) Roots are real, distinct and negative (since $\alpha\beta > 0 \Rightarrow$ same signs and $\alpha + \beta < 0 \Rightarrow$ both < 0)

   (ii) $\alpha = -\beta$

   (iii) One of the roots is zeros and the other is $-\frac{b}{a}$.

   (iv) The roots are real and of opposite signs.

5. Let $az^2 + bz + c = 0$ have roots α and 2α.

   Sum of roots $\alpha + 2\alpha = 3\alpha = -\frac{b}{a}$ so $\alpha = -\frac{b}{3a}$

   Product of roots $\alpha \times 2\alpha = 2\alpha^2 = \frac{c}{a}$ so

   $2 \times \left(-\frac{b}{3a}\right)^2 = \frac{c}{a}$

   Then $2b^2 = 9ac$ as required.

6. (i) $az^2 + bkz + ck^2 = 0$

   (ii) $az^2 + (b - 2ka)z + (k^2a - kb + c) = 0$

7. (ii) $z^2 - (5 + 2i) + (9 + 7i) = 0$

Exercise 3.2 (Page 57)

1. (i) $-\frac{3}{2}$

   (ii) $-\frac{1}{2}$

   (iii) $-\frac{7}{2}$

2. (i) $z^3 - 7z^2 + 14z - 8 = 0$

   (ii) $z^3 - 3z^2 - 4z + 12 = 0$

(iii) $2z^3 + 7z^2 + 6z = 0$

(iv) $2z^3 - 13z^2 + 28z - 20 = 0$

(v) $z^3 - 19z - 30 = 0$

(vi) $z^3 - 5z^2 + 9z - 5 = 0$

3   (i) $z = 2, 5, 8$

(ii) $z = -\frac{2}{3}, \frac{2}{3}, 2$

(iii) $z = 2 - 2\sqrt{3}, 2, 2 + 2\sqrt{3}$

(iv) $z = \frac{2}{3}, \frac{7}{6}, \frac{5}{3}$

4   (i) $z = w - 3$

(ii) $(w - 3)^3 + (w - 3)^2 + 2(w - 3) - 3 = 0$

(iii) $w^3 - 8w^2 + 23w - 27 = 0$

(iv) $\alpha + 3, \beta + 3, \gamma + 3$

5   $w^3 - 4w^2 + 4w - 24 = 0$

6   (i) $2w^3 - 16w^2 + 37w - 27 = 0$

(ii) $2w^3 + 24w^2 + 45w + 37 = 0$

7   The roots are $\frac{3}{2}, 2, \frac{5}{2}$ $k = \frac{47}{2}$

8   $z = \frac{1}{4}, \frac{1}{2}, -\frac{3}{4}$

9   $\alpha = -1, p = 7, q = 8$ or $\alpha = p = q = 0$

10 Roots are $-p$ and $\pm\sqrt{-q}$ (note $\pm\sqrt{-q}$ is not necessarily imaginary, since q is not necessarily >0)

11 (i) $p = -8\left(\alpha + \frac{1}{2\alpha} + \beta\right)$

$q = 8\left(\frac{1}{2} + \alpha\beta + \frac{\beta}{2\alpha}\right)$

$r = -4\beta$

(iii) $r = 9; x = 1, \frac{1}{2}, -\frac{9}{4}$

$r = -6; x = -2, -\frac{1}{4}, \frac{3}{2}$

12 $z = \frac{3}{7}, \frac{7}{3}, -2$

13 $ac^3 = b^3 d$

$z = \frac{1}{2}, \frac{3}{2}, \frac{9}{2}$

Exercise 3.3 (Page 60)

1   (i) $-\frac{3}{2}$

(ii) 3

(iii) $\frac{5}{2}$

(iv) 2

2   (i) $z^4 - 6z^3 + 7z^2 + 6z - 8 = 0$

(ii) $4z^4 + 20z^3 + z^2 - 60z = 0$

(iii) $4z^4 + 12z^3 - 27z^2 - 54z + 81 = 0$

(iv) $z^4 - 5z^2 + 10z - 6 = 0$

3   (i) $z^4 + 4z^3 - 6z^2 - 4z + 48 = 0$

(ii) $2z^4 + 12z^3 + 21z^2 + 13z + 8 = 0$

4   (i) Let $w = x + 1$ then $x = w - 1$
new quartic: $x^4 - 6x^2 + 9$

(ii) Solutions to new quartic are $x = \pm\sqrt{3}$ (each one repeated), solutions to original quartic are therefore: $\alpha = \beta = \sqrt{3} - 1$ and $\gamma = \delta = -\sqrt{3} - 1$.

5   (i) $\alpha = -1, \ \beta = \sqrt{3}$

(ii) $p = 4$ and $q = -9$

(iii) Use substitution $y = x - 3\alpha$ (i.e. $y = x + 3$ then $x = y - 3$) and $y^3 - 8y^2 + 18y - 12 = 0$

6   (i)

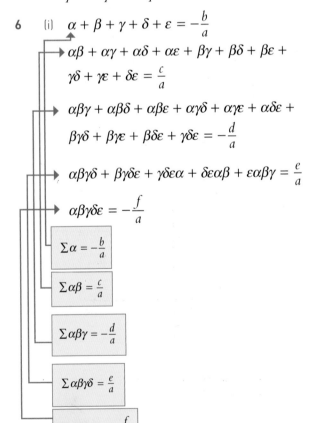

$\alpha + \beta + \gamma + \delta + \varepsilon = -\dfrac{b}{a}$

$\alpha\beta + \alpha\gamma + \alpha\delta + \alpha\varepsilon + \beta\gamma + \beta\delta + \beta\varepsilon + \gamma\delta + \gamma\varepsilon + \delta\varepsilon = \dfrac{c}{a}$

$\alpha\beta\gamma + \alpha\beta\delta + \alpha\beta\varepsilon + \alpha\gamma\delta + \alpha\gamma\varepsilon + \alpha\delta\varepsilon + \beta\gamma\delta + \beta\gamma\varepsilon + \beta\delta\varepsilon + \gamma\delta\varepsilon = -\dfrac{d}{a}$

$\alpha\beta\gamma\delta + \beta\gamma\delta\varepsilon + \gamma\delta\varepsilon\alpha + \delta\varepsilon\alpha\beta + \varepsilon\alpha\beta\gamma = \dfrac{e}{a}$

$\alpha\beta\gamma\delta\varepsilon = -\dfrac{f}{a}$

$\Sigma\alpha = -\dfrac{b}{a}$

$\Sigma\alpha\beta = \dfrac{c}{a}$

$\Sigma\alpha\beta\gamma = -\dfrac{d}{a}$

$\Sigma\alpha\beta\gamma\delta = \dfrac{e}{a}$

$\Sigma\alpha\beta\gamma\delta\varepsilon = -\dfrac{f}{a}$

Exercise 3.4 (Page 63)

1    $4 + 5i$ is the other root.

     The equation is $z^2 - 8z + 41 = 0$.

2    $2 - i, -3$

3    $7, 4 \pm 2i$

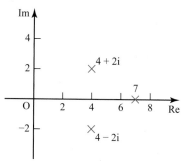

4   (i) $z = -3$

     (ii) $z = -3, \dfrac{5}{2} \pm \dfrac{\sqrt{11}}{2} i$

5    $k = 36$, other roots are $-\dfrac{3}{2} \pm \dfrac{3\sqrt{3}}{2} i$

6    $p = 4, q = -10$, other roots are $1 + i$ and -6

7    $z^3 - z - 6 = 0$

8    $z = 3 \pm 2i, 2 \pm i$

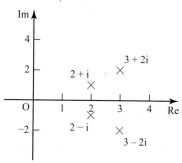

9   (i) $w^2 = -2i, w^3 = -2 - 2i, w^4 = -4$

     (ii) $p = -4, q = 2$

     (iii) $z = -4, -1, 1 \pm i$

10   (i) $z = \pm 3, \pm 3i$

     (ii)

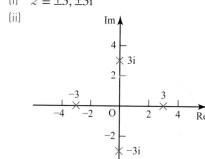

11   (i) $\alpha^2 = -3 - 4i, \alpha^3 = 11 - 2i$

     (ii) $z = -1 - 2i, -5$

(iii)

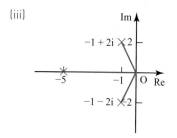

12   A false, B true, C true, D true

13    $a = 2, b = 2, z = -2 \pm i, 1 \pm 2i$

14    $z = \pm 3i, 4 \pm \sqrt{5}$

15   (i) $\alpha^2 = -8 - 6i, \alpha^3 = 26 - 18i$

     (ii) $\mu = 20$

     (iii) $z = -1 \pm 3i, -\dfrac{2}{3}$

     (iv)

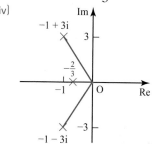

16    $a = 1, b = -9, c = 44, d = -174, e = 448,$
      $f = -480$

Chapter 4

Discussion point (Page 67)

You might have said, for example, that the sequence is increasing or that each term is 3 more than the previous terms.

Exercise 4.1 (Page 70)

1   (i) $6, 11, 16, 21, 26$

        Increasing by 5 for each term

     (ii) $-3, -9, -15, -21, -27$

        Decreasing by 6 for each term

     (iii) $8, 16, 32, 64, 128$

        Doubling for each term

     (iv) $8, 12, 8, 12, 8$

        Oscillating

     (v) $2, 5, 11, 23, 47$

        Increasing

     (vi) $5, \dfrac{5}{2}, \dfrac{5}{3}, \dfrac{5}{4}, 1$

        Decreasing, converging to zero

2 (i) 21, 25, 29, 33

(ii) $u_1 = 1$, $u_{r+1} = u_r + 4$

(iii) $u_r = 4r - 3$

3 (i) (a) 0, −2, −4, −6

(b) $u_1 = 10$, $u_{r+1} = u_r - 2$

(c) $u_r = 12 - 2r$

(d) −28

(ii) (a) 32, 64, 128, 256

(b) $u_1 = 1$, $u_{r+1} = 2u_r$

(c) $u_r = 2^{r-1}$

(d) 524 288

(iii) (a) 31 250, 156 250, 781 250, 3 906 250

(b) $u_1 = 50$, $u_{r+1} = 5u_r$

(c) $u_r = 10 \times 5^r$

(d) 9.54×10^{14} (3 s.f.)

4 (i) 25

(ii) −150

(iii) 363

(iv) −7.5

5 (i) $\displaystyle\sum_{r=1}^{7}(56 - 6r)$

(ii) 224

6 2500

7 (i) −5, 5, −5, 5, −5, 5

Oscillating

(ii) (a) 0 (b) −5

(iii) $-\dfrac{5}{2} + \dfrac{5}{2}(-1)^n$

8 (i) 0, 100, 2, 102, 4, 104

Even terms start from 100 and increase by 2, odd terms start from 0 and increase by 2.

(ii) 201

(iii) 102

9 749 cm

10 $\dfrac{1}{2}n(n^3 + 1)$

11 10, 5, 16, 8, 4 (This will reach 1 at c_7 and then repeat the cycle 4, 2, 1)

Exercise 4.2 (Page 73)

1 (i) 1, 3, 5   (ii) n^2

2 (i) 4, 14, 30   (ii) $n(n + 1)^2$

3 (i) 2, 12, 36   (ii) $\dfrac{1}{12}n(n + 1)(3n + 1)(n + 2)$

4 n^4

5 $\dfrac{1}{3}n(n + 1)(n + 2)$

6 $\dfrac{1}{4}n(n + 1)(n + 2)(n + 3)$

7 $\dfrac{1}{2}n(3n + 1)$

8 $n^2(4n + 1)(5n + 2)$

9 (ii) 7 layers, 125 left over

10 (i) £227.50   (ii) $\dfrac{1}{24}n(35(n + 1) + 30I)$

Activity 4.1 (Page 75)

$$\frac{1}{1 \times 2} = \frac{1}{2}$$

$$\frac{1}{1 \times 2} + \frac{1}{2 \times 3} = \frac{2}{3}$$

$$\frac{1}{1 \times 2} + \frac{1}{2 \times 3} + \frac{1}{3 \times 4} = \frac{3}{4}$$

$$\frac{1}{1 \times 2} + \frac{1}{2 \times 3} + \frac{1}{3 \times 4} + \frac{1}{4 \times 5} = \frac{4}{5}$$

Activity 4.2 (Page 78)

(i) Assume true for $n = k$, so

$$2 + 4 + 6 + \ \dots \ + 2k = \left(k + \frac{1}{2}\right)^2.$$

For $n = k + 1$,

$2 + 4 + 6 + \ \dots \ +$

$$2k + 2(k + 1) = \left(k + \frac{1}{2}\right)^2 + 2(k + 1)$$

$$= k^2 + k + \frac{1}{4} + 2k + 2$$

$$= k^2 + 3k + \frac{9}{4}$$

$$= \left(k + \frac{3}{2}\right)^2$$

$$= \left(k + 1 + \frac{1}{2}\right)^2$$

It is not true for $n = 1$.

(ii) It breaks down at the inductive step.

Practice Questions 1 (Page 84)

1 (i) $\begin{pmatrix} 7 & 5 \\ 2 & -5 \end{pmatrix}$ [3]

(ii) $\begin{pmatrix} 7 & 4 \\ 6 & 7 \end{pmatrix} + \begin{pmatrix} 4 & 8 \\ 12 & 4 \end{pmatrix} = \begin{pmatrix} 11 & 12 \\ 18 & 11 \end{pmatrix}$

[1], [1], [1]

2 (i) Points plotted at $1 + 2i$, $-3 + 4i$, $4i$, $\dfrac{2}{5}$

[1], [1], [1], [2]

(ii) w^2, $w - w^*$ [1]

3 *Either:*

Cubic has real coefficients [1]

so $3 - i$ a root [1]

Sum of $3 + i$ and $3 - i$ is 6; sum of all 3 roots is 9 [1]

so real root is 3 [1]

Or:

$z = 3$ is a root by trying factors of 30 [1]

From factor theorem $(z - 3)$ a factor of the cubic [1]

$z^3 - 9z^2 + 28z - 30 = (z - 3)$
$(z^2 - 6z + 10)$ [1]

Roots of quadratic are $3 + i$, $3 - i$ [1]

4 (i) $= \dfrac{-6 \pm \sqrt{36 - 20}}{2(2 + i)}$ [1]

$= \dfrac{-5}{2 + i}$ or $\dfrac{-1}{2 + i}$ [1]

$= \dfrac{-5(2 - i)}{(2 + i)(2 - i)}$ or $\dfrac{-(2 - i)}{(2 + i)(2 - i)}$ [1]

$= -(2 - i)$ or $= -\dfrac{1}{5}(2 - i)$: both solutions are in the form $\lambda(2 - i)$ with $\lambda = -1$ and $\lambda = -\dfrac{1}{5}$ [1]

(ii) By substituting the roots into the equation. [1]

5 (i) $\alpha + 1, \beta + 1, \gamma + 1$ satisfy
$(\gamma - 1)^3 + 3(\gamma - 1)^2 - 6(\gamma - 1) - 8 = 0$ [1]
$\gamma^3 - 3\gamma^2 + 3\gamma - 1 + 3\gamma^2 - 6\gamma$
$\quad + 3 - 6\gamma + 6 - 8 = 0$ [1], [1]
$\gamma^3 - 9\gamma = 0$ [1]

(ii) $\gamma(\gamma^2 - 9) = 0$ [1]
$\gamma(\gamma - 3)(\gamma + 3) = 0$ [1]
$\gamma = 0, 3, -3$ [1]

(iii) $x = -1, 2, -4$ [2]

6 (i) Diagram or calculation showing image of shape/points

Rotation 90°… [1]

… about $(0, 0)$, anticlockwise [1]

(ii) Rotation 45° anticlockwise about $(0, 0)$, when repeated, gives transformation corresponding to **B**. [1]

Diagram showing, for example, unit square or unit vectors rotated by 45°. [1]

$\begin{pmatrix} \dfrac{1}{\sqrt{2}} & -\dfrac{1}{\sqrt{2}} \\ \dfrac{1}{\sqrt{2}} & \dfrac{1}{\sqrt{2}} \end{pmatrix}$ [1]

7 $\delta + (\delta + 1) = -\dfrac{b}{a}$ $b^2 = a^2(\delta + (\delta + 1))^2$
$= a^2(\delta^2 + 2\delta(\delta + 1) + (\delta + 1)^2)$
$= a^2(4\delta^2 + 4\delta + 1)$ [1]

$\delta(\delta + 1) = \dfrac{c}{a} \Rightarrow ac = a^2(\delta(\delta + 1))$
$= a^2(\delta^2 + \delta)$ [1]

LHS $= b^2 - 4ac$
$= a^2(4\delta^2 + 4\delta + 1 - 4(\delta^2 + \delta))$ [1]
$= a^2(1) = a^2$ [1]
$=$ RHS(complete argument, well set out) [1]

8 (i) $3, 6, 11, 20, 37$ [1]

(ii) To prove $u_n = 2^n + n$

When $n = 1$, LHS $= 3$, RHS $= 2^1 + 1$
$\quad = 2 + 1 = 3$ [1]

So it is true for $n = 1$

Assume it is true for $n = k$, so:
$u_k = 2^k + k$ [1]

Want to show that $u_{k+1} = 2^{k+1} + k + 1$.
$u_{k+1} = 2u_k - k + 1 = 2(2^k + k) - k + 1$ [1]
$= 2^{k+1} + 2k - k + 1 = 2^{k+1} + k + 1$ as required.

So, if the result is true for $n = k$ then it is true for $n = k + 1$ [1]

Since it is true for $n = 1$, by induction it is true for all positive integers n. [1]

9 (i) Calculations or image correct for three points. [1]

Totally correct plot of $(0, 0)$ $(-0.6, 0.8)$ $(0.2, 1.4)$ $(0.8, 0.6)$ [1]

(ii) $\begin{pmatrix} -\dfrac{3}{5} & \dfrac{4}{5} \\ \dfrac{4}{5} & \dfrac{3}{5} \end{pmatrix} \begin{pmatrix} x \\ y \end{pmatrix} = \begin{pmatrix} x \\ y \end{pmatrix}$

$-\dfrac{3}{5}x + \dfrac{4}{5}y = x$ [1]

$\dfrac{4}{5}x + \dfrac{3}{5}y = y$

$\left.\begin{matrix} y = 2x \\ y = 2x \end{matrix}\right\}$ from both equations [1]

$y = 2x$ is equation of line of invariant points. [1]

(iii) Perpendicular line to this, through origin, is $y = -\dfrac{1}{2}x$.

$$\begin{pmatrix} -\dfrac{3}{5} & \dfrac{4}{5} \\ \dfrac{4}{5} & \dfrac{3}{5} \end{pmatrix} \begin{pmatrix} x \\ -\dfrac{1}{2}x \end{pmatrix} = \begin{pmatrix} -\dfrac{3}{5}x - \dfrac{2}{5}x \\ \dfrac{4}{5}x - \dfrac{3}{10}x \end{pmatrix}$$

$$= \begin{pmatrix} -x \\ \dfrac{1}{2}x \end{pmatrix}$$

$$= -\begin{pmatrix} x \\ -\dfrac{1}{2}x \end{pmatrix} \qquad [1]$$

So $y = -\dfrac{1}{2}x$ is an invariant line, and is perpendicular to line of invariant points, and both go through the origin. [1]

(iv) Two points marked, where image of unit square intersects unit square, at $(0, 0)$ and $(0.5, 1)$. [1], [1]

10 (i) To prove $f(n) = \dfrac{n}{3} + \dfrac{n^2}{2} + \dfrac{n^3}{6}$ is integer-valued.
$f(1) = \dfrac{1}{3} + \dfrac{1}{2} + \dfrac{1}{6} = 1$, which is integer-valued [1]
So it is true for $n = 1$.
Assume it is true for $n = k$,
so $f(k) = \dfrac{k}{3} + \dfrac{k^2}{2} + \dfrac{k^3}{6}$ is integer-valued. [1]
Want to show that $f(k + 1)$ is integer-valued
$f(k + 1) = \dfrac{k + 1}{3} + \dfrac{(k + 1)^2}{2} + \dfrac{(k + 1)^3}{6}$
$= \dfrac{k + 1}{3} + \dfrac{k^2 + 2k + 1}{2}$
$\qquad + \dfrac{k^3 + 3k^2 + 3k + 1}{6}$
$= \left(\dfrac{k}{3} + \dfrac{k^2}{2} + \dfrac{k^3}{6} \right) + \left(\dfrac{1}{3} + \dfrac{1}{2} + \dfrac{1}{6} \right)$
$\qquad + \dfrac{2k}{2} + \dfrac{3k^2}{6} + \dfrac{3k}{6}$ [1]
$= f(k) + 1 + k + \dfrac{1}{2}k(k + 1)$
$k(k + 1)$ is a product of two consecutive integers, so one must be even and
$\dfrac{1}{2}k(k + 1)$ is an integer, so $f(k + 1) =$ integer $+ 1 +$ integer $+$ integer, which is an integer, as required.

So, if the result is true for $n = k$ then it is true for $n = k + 1$. [1]
Since it is true for $n = 1$, by induction it is true for all positive integers n. [1]

(ii) $\dfrac{n}{3} + \dfrac{n^2}{2} + \dfrac{n^3}{6} = \dfrac{2n + 3n^2 + n^3}{6}$ [1]
$= \dfrac{n(n^2 + 3n + 2)}{6} = \dfrac{n(n + 1)(n + 2)}{6}$ [1]
$n(n + 1)(n + 2)$ is a product of three consecutive integers, so one must be a multiple of 3 and at least one must be even, hence $n(n + 1)(n + 2)$ is a multiple of 6. [1]
So $\dfrac{n(n + 1)(n + 2)}{6}$ is integer-valued.

11 (i) To prove $\begin{pmatrix} 2 & -1 \\ 0 & 1 \end{pmatrix}^n = \begin{pmatrix} 2^n & 1 - 2^n \\ 0 & 1 \end{pmatrix}$

When $n = 1$, LHS $= \begin{pmatrix} 2 & -1 \\ 0 & 1 \end{pmatrix}$ and

RHS $= \begin{pmatrix} 2 & 1 - 2 \\ 0 & 1 \end{pmatrix} = \begin{pmatrix} 2 & -1 \\ 0 & 1 \end{pmatrix}$ [1]

So it is true for $n = 1$.
Assume it is true for $n = k$,

so $\begin{pmatrix} 2 & -1 \\ 0 & 1 \end{pmatrix}^k = \begin{pmatrix} 2^k & 1 - 2^k \\ 0 & 1 \end{pmatrix}$ [1]

Then:
$$\begin{pmatrix} 2 & -1 \\ 0 & 1 \end{pmatrix}^{k+1} = \begin{pmatrix} 2 & -1 \\ 0 & 1 \end{pmatrix}^k \begin{pmatrix} 2 & -1 \\ 0 & 1 \end{pmatrix}$$

$$= \begin{pmatrix} 2^k & 1 - 2^k \\ 0 & 1 \end{pmatrix} \begin{pmatrix} 2 & -1 \\ 0 & 1 \end{pmatrix} \qquad [1]$$

$$= \begin{pmatrix} (2^k)(2) + 0 & (2^k)(-1) + (1 - 2^k)(1) \\ 0 + 0 & 0 + 1 \end{pmatrix}$$

$$= \begin{pmatrix} 2^{k+1} & 1 - 2^{k+1} \\ 0 & 1 \end{pmatrix}$$

So, if the result is true for $n = k$ then it is true for $n = k + 1$. [1]
Since it is true for $n = 1$, by induction it is true for all positive integers n. [1]

(ii) To prove $5^n - 4n - 1$ is divisible by 8:
When $n = 1$, $5^1 - 4(1) - 1 = 5 - 4 - 1 = 0$, which is divisible by 8 [1]
So it is true for $n = 1$.
Assume it is true for $n = k$,
so $5^k - 4k - 1$ is a multiple of 8. [1]

Then one of the following:

(I) $f(n) = 5^n - 4n - 1$
$f(k+1) - f(k) = (5^{k+1} - 4(k + 1) - 1) - (5^k - 4k - 1)$
$= 5(5^k) - 5^k - 4 = 4(5^k - 1)$
$= 4(5 - 1)(5^{k-1} + 5^{k-2} + \dots + 5 + 1)$ [1]

So $f(k + 1) - f(k)$ is divisible by 16 and hence $f(k + 1)$ is divisible by 8.

(II) $f(n) = 5^n - 4n - 1$
$f(k + 1) - f(k) = (5^{k+1} - 4(k + 1) - 1) - (5^k - 4k - 1)$
$= 5(5^k) - 5^k - 4 = 4(5^k - 1) = 4(f(k)$
$+ 4k) = 4(\text{multiple of } 8 + 4k)$ [1]

So $f(k + 1) - f(k)$ is divisible by 16 and hence $f(k + 1)$ is divisible by 8.

(III) $f(n) = 5^n - 4n - 1$
$f(k + 1) - f(k) = (5^{k+1} - 4(k + 1) - 1) - (5^k - 4k - 1)$
$= 5(5^k) - 5^k - 4 = 4(5^k - 1) = 4(f(k) + 4k)$
So $f(k + 1) = 4f(k) + 16k$ [1]

So $f(k + 1)$ is divisible by 8.

(IV) $5^{k+1} - 4(k + 1) - 1 = 5(5^k - 4k - 1) + 20k$
$+ 5 - 4k - 4 - 1$
$= 5(5^k - 4k - 1) + 16k$ [1]
So $5^{k+1} - 4(k + 1) - 1$ is divisible by 8.

So, if the result is true for $n = k$ then it is true for $n = k + 1$ [1]

Since it is true for $n = 1$, by induction it is true for all positive integers n. [1]

Chapter 5

Discussion point (Page 86)

Find the volume of a cone with radius 2 and height 2, and subtract the volume of a cone with radius 1 and height 1.

Volumes of some other solids of revolution can be found using formulae, but many cannot.

Exercise 5.1 (Page 91)

1 For example; ball, tin of soup, bottle of wine, roll of sticky tape, dinner plate

2 (i) cone, radius 18 and height 6
   (ii) 648π cubic units
   (iii) cone, radius 2, height 6
   (iv) 8π cubic units

3 $\frac{1}{2}\pi$

4 (i)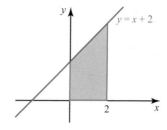

   (ii) $\frac{56}{3}\pi$ cubic units

5 (i)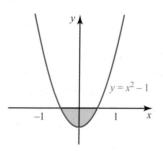

   (ii) $\frac{16}{15}\pi$ cubic units

6 (i)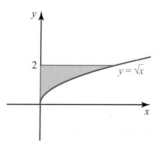

   (ii) 6.4π cubic units

7 (i)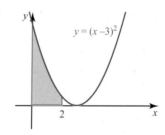

   (ii) $\frac{26}{3}$

   (iii) $\frac{242}{5}\pi$ cubic units

8 (i)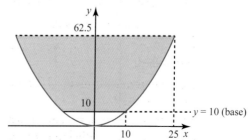

(ii) 45.9 litres

9 (i)

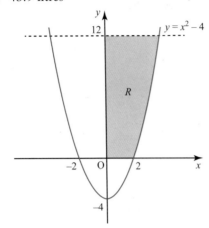

(ii) $\displaystyle\int_0^{12} \pi(y+4)\,dy$ (iii) 3 litres

10 (i) A$(-3, 4)$ B$(3, 4)$ (ii) 36π cubic units

Chapter 6

Discussion point (Page 96)

An internet search should give you lots of information about the Mandelbrot set. The Mandelbrot set consists of all complex numbers, c, for which the complex number sequence defined by $z_1 = c$, $z_{n+1} = z_n^2 + c$ does not escape to infinity.

Discussion point (Page 98)

It is not true that $\arg(z)$ is given by $\arctan\left(\dfrac{y}{x}\right)$. For example the complex number $-1 + i$ has argument $\dfrac{3\pi}{4}$ but $\arctan\left(\dfrac{1}{-1}\right) = -\dfrac{\pi}{4}$. A diagram is needed to ensure the correct angle is calculated.

Activity 6.1 (Page 101)

	$\dfrac{\pi}{6}$	$\dfrac{\pi}{4}$	$\dfrac{\pi}{3}$
sin	$\dfrac{1}{2}$	$\dfrac{1}{\sqrt{2}}$	$\dfrac{\sqrt{3}}{2}$
cos	$\dfrac{\sqrt{3}}{2}$	$\dfrac{1}{\sqrt{2}}$	$\dfrac{1}{2}$
tan	$\dfrac{1}{\sqrt{3}}$	1	$\sqrt{3}$

Exercise 6.1 (Page 103)

1   $z_1 = 4$ or $4(\cos 0 + i\sin 0)$

    $z_2 = -2 + 4i$ or $2\sqrt{5}(\cos 2.03 + i\sin 2.03)$

    $z_3 = 1 - 3i$ or

    $\sqrt{10}\left(\cos(-1.25) + i\sin(-1.25)\right)$

2   (i) $|z| = \sqrt{13}$ $\arg(z) = 0.588$

    (ii) $|z| = \sqrt{29}$ $\arg(z) = 2.76$

    (iii) $|z| = \sqrt{13}$ $\arg(z) = -2.55$

    (iv) $|z| = \sqrt{29}$ $\arg(z) = -1.19$

3   $|z_1| = \sqrt{13}$ $\arg(z_1) = 0.588$

    $|z_2| = \sqrt{13}$ $\arg(z_2) = -0.588$

    $|z_3| = \sqrt{13}$ $\arg(z_3) = -2.16$

    $|z_4| = \sqrt{13}$ $\arg(z_4) = 2.16$

    $z_1 \to z_2$ Reflection in real axis

    $z_1 \to z_3$ Reflection in the line $y = -x$

    $z_1 \to z_4$ Rotation of $90°$ anticlockwise about the origin

4   (i) $-4i$

    (ii) $-\dfrac{7}{\sqrt{2}} + \dfrac{7}{\sqrt{2}}i$

    (iii) $-\dfrac{3\sqrt{3}}{2} + \dfrac{3}{2}i$

    (iv) $\dfrac{5\sqrt{3}}{2} - \dfrac{5}{2}i$

5   (i) $1(\cos 0 + i\sin 0)$

    (ii) $2(\cos \pi + i\sin \pi)$

    (iii) $3\left(\cos\left(\dfrac{\pi}{2}\right) + i\sin\left(\dfrac{\pi}{2}\right)\right)$

    (iv) $4\left(\cos\left(-\dfrac{\pi}{2}\right) + i\sin\left(-\dfrac{\pi}{2}\right)\right)$

6   (i) $\sqrt{2}\left(\cos\dfrac{\pi}{4} + i\sin\dfrac{\pi}{4}\right)$

    (ii) $\sqrt{2}\left(\cos\dfrac{3\pi}{4} + i\sin\dfrac{3\pi}{4}\right)$

    (iii) $\sqrt{2}\left(\cos\left(\dfrac{-3\pi}{4}\right) + i\sin\left(\dfrac{-3\pi}{4}\right)\right)$

    (iv) $\sqrt{2}\left(\cos\left(\dfrac{-\pi}{4}\right) + i\sin\left(\dfrac{-\pi}{4}\right)\right)$

7   (i) $12\left(\cos\dfrac{\pi}{6} + i\sin\dfrac{\pi}{6}\right)$

    (ii) $5\left(\cos(-0.927) + i\sin(-0.927)\right)$

(iii) $13\left(\cos 2.75 + i \sin 2.75\right)$

(iv) $\sqrt{65}\left(\cos 1.05 + i \sin 1.05\right)$

(v) $\sqrt{12013}\left(\cos\left(-2.13\right) + i \sin\left(-2.13\right)\right)$

8 (i) $\frac{1}{5}\sqrt{10}\left(\cos 0.322 + i \sin 0.322\right)$

(ii) $\frac{\sqrt{130}}{10}\left(\cos\left(-0.266\right) + i \sin\left(-0.266\right)\right)$

(iii) $\frac{\sqrt{290}}{10}\left(\cos\left(-1.63\right) + i \sin\left(-1.63\right)\right)$

9 (i) $z = 2i$ or $0 + 2i$

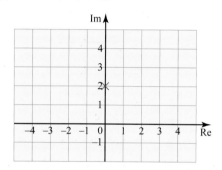

(ii) $z = \frac{3}{2} + \frac{3\sqrt{3}}{2} i$

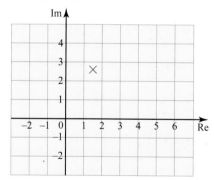

(iii) $z = -\frac{7\sqrt{3}}{2} + \frac{7}{2} i$

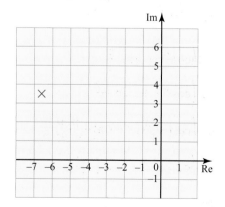

(iv) $z = \frac{1}{\sqrt{2}} - \frac{1}{\sqrt{2}} i$

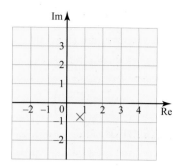

(v) $z = -\frac{5}{2} - \frac{5\sqrt{3}}{2} i$

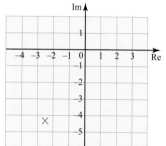

(vi) $z = -2.50 - 5.46i$

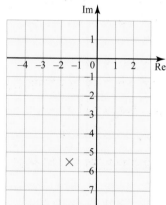

10 (i) $-(\pi - \alpha)$ or $\alpha - \pi$ (ii) $-\alpha$

(iii) $\pi - \alpha$ (iv) $\frac{\pi}{2} - \alpha$

(v) $\frac{\pi}{2} + \alpha$

11 (i) $\left|z_1\right| = 5$ $\arg\left(z_1\right) = 0.927$

$\left|z_2\right| = \sqrt{2}$ $\arg\left(z_2\right) = \frac{3\pi}{4}$

(ii) (a) $z_1 z_2 = -7 - i$ $\frac{z_1}{z_2} = \frac{1}{2} - \frac{7}{2} i$

(b) $\left|z_1 z_2\right| = 5\sqrt{2}$ $\arg\left(z_1 z_2\right) = -3.00$

$\left|\frac{z_1}{z_2}\right| = \frac{5\sqrt{2}}{2}$ $\arg\left(\frac{z_1}{z_2}\right) = -1.43$

(iii) $|z_1 z_2| = |z_1||z_2|$ and $\left|\dfrac{z_1}{z_2}\right| = \dfrac{|z_1|}{|z_2|}$

$\arg(z_1) + \arg(z_2) = 3.28$ which is greater than π, but is equivalent to -3.00

i.e. $\arg(z_1) + \arg(z_2) = \arg(z_1 z_2)$

$\arg(z_1) - \arg(z_2) = \arg\left(\dfrac{z_1}{z_2}\right)$

Activity 6.3 (Page 105)

(i) Rotation of 90° anticlockwise about the origin

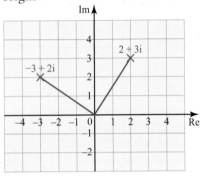

(ii) Rotation of 90° anticlockwise about the origin and enlargement of scale factor 2

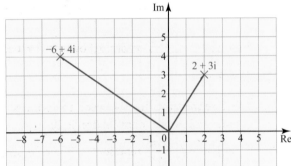

(iii) Rotation of $\dfrac{\pi}{4}$ anticlockwise and enlargement of scale factor $\sqrt{2}$

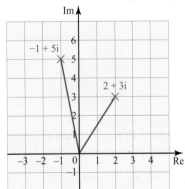

Exercise 6.2 (Page 107)

1 (i) $|w| = \sqrt{2}$ $\arg(w) = \dfrac{\pi}{4}$

     $|z| = 2$ $\arg(z) = -\dfrac{\pi}{3}$

(ii) (a) $|wz| = 2\sqrt{2}$ $\arg(wz) = -\dfrac{\pi}{12}$

     (b) $\left|\dfrac{w}{z}\right| = \dfrac{1}{\sqrt{2}}$ $\arg\left(\dfrac{w}{z}\right) = \dfrac{7\pi}{12}$

(iii)

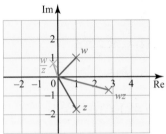

2 (i) $6\left(\cos\dfrac{7\pi}{12} + i\sin\dfrac{7\pi}{12}\right)$

(ii) $\dfrac{3}{2}\left(\cos\left(\dfrac{\pi}{12}\right) + i\sin\left(\dfrac{\pi}{12}\right)\right)$

(iii) $\dfrac{2}{3}\left(\cos\left(-\dfrac{\pi}{12}\right) + i\sin\left(-\dfrac{\pi}{12}\right)\right)$

(iv) $\dfrac{1}{2}\left(\cos\left(-\dfrac{\pi}{4}\right) + i\sin\left(-\dfrac{\pi}{4}\right)\right)$

3 (i) $\dfrac{2\pi}{3}$ (ii) 6

(iii) $\dfrac{\pi}{2}$ (iv) 108

4 (i) $4\left(\cos\left(-\dfrac{\pi}{2}\right) + i\sin\left(-\dfrac{\pi}{2}\right)\right)$

(ii) $7776\left(\cos\left(\dfrac{5\pi}{6}\right) + i\sin\left(\dfrac{5\pi}{6}\right)\right)$

(iii) $10368\left(\cos\left(-\dfrac{\pi}{12}\right) + i\sin\left(-\dfrac{\pi}{12}\right)\right)$

(iv) $30\left(\cos\left(\dfrac{2\pi}{3}\right) + i\sin\left(\dfrac{2\pi}{3}\right)\right)$

(v) $2\sqrt{2}\left(\cos 0 + i\sin 0\right)$

5 (i) Multiplication scale factor $\dfrac{\sqrt{377}}{13}$, angle of rotation -1.36 radians (i.e. 1.36 radians clockwise)

(ii) Multiplication scale factor $\dfrac{3}{\sqrt{17}}$, angle of rotation -1.33 radians (i.e. 1.33 radians clockwise)

6 $\arg\left(\dfrac{w}{z}\right) = \arg(w) - \arg(z) \Rightarrow \arg\left(\dfrac{1}{z}\right)$

$= \arg(1) - \arg(z) = 0 - \arg(z) = -\arg(z)$

The exceptions are complex numbers for which both $\operatorname{Im}(z) = 0$ and $\operatorname{Re}(z) \leqslant 0$ since $-180 < \arg(z) \leqslant 180$

7 (i) Real part $= \dfrac{-1 + \sqrt{3}}{4}$

Imaginary part $= \dfrac{1 + \sqrt{3}}{4}$

(ii) $\sqrt{2}\left(\cos\left(\dfrac{3\pi}{4}\right) + i\sin\left(\dfrac{3\pi}{4}\right)\right)$

$2\left(\cos\left(\dfrac{\pi}{3}\right) + i\sin\left(\dfrac{\pi}{3}\right)\right)$

(iii)

$\dfrac{-1 + i}{1 + \sqrt{3}i} = \dfrac{\sqrt{2}\left(\cos\left(\dfrac{3\pi}{4}\right) + i\sin\left(\dfrac{3\pi}{4}\right)\right)}{2\left(\cos\left(\dfrac{\pi}{3}\right) + i\sin\left(\dfrac{\pi}{3}\right)\right)}$

$= \dfrac{1}{\sqrt{2}}\left(\cos\left(\dfrac{5\pi}{12}\right) + i\sin\left(\dfrac{5\pi}{12}\right)\right)$

$= \dfrac{-1 + \sqrt{3}}{4} + \dfrac{1 + \sqrt{3}}{4}i$

$\Rightarrow \cos\left(\dfrac{5\pi}{12}\right) = \dfrac{\sqrt{3} - 1}{2\sqrt{2}} \qquad \sin\left(\dfrac{5\pi}{12}\right) = \dfrac{\sqrt{3} + 1}{2\sqrt{2}}$

8 For the complex numbers
$w = r_1\left(\cos\theta_1 + i\sin\theta_1\right)$ and
$z = r_2\left(\cos\theta_2 + i\sin\theta_2\right)$ we have proven that
$wz = r_1 r_2\left[(\cos(\theta_1 + \theta_2) + i\sin(\theta_1 + \theta_2)\right]$
So,

$wzp = r_1 r_2[(\cos(\theta_1 + \theta_2) + i\sin(\theta_1 + \theta_2)] \times r_3\left(\cos\theta_3 + i\sin\theta_3\right)$

$= r_1 r_2 r_3[\cos(\theta_1 + \theta_2)\cos\theta_3 + i\sin(\theta_1 + \theta_2)\sin\theta_3 +$

$\qquad i\sin(\theta_1 + \theta_2)\cos\theta_3 + i^2\sin(\theta_1 + \theta_2)\sin\theta_3]$

$= r_1 r_2 r_3\{[\cos(\theta_1 + \theta_2)\cos\theta_3 - \sin(\theta_1 + \theta_2)\sin\theta_3]$

$\qquad + i[\cos(\theta_1 + \theta_2)\sin\theta_3 + \sin(\theta_1 + \theta_2)\cos\theta_3]\}$

$= r_1 r_2 r_3\{\cos[(\theta_1 + \theta_2) + \theta_3] + i\sin[(\theta_1 + \theta_2) + \theta_3]\}$

Therefore, $|wzp| = |w|\,|z|\,|p|$ and

$\arg(wzp) = \arg(z) + \arg(w) + \arg(p)$.

Activity 6.4 (Page 112)

(i)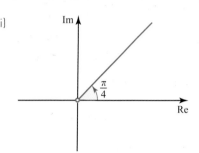

$\arg(z) = \dfrac{\pi}{4}$ represents a half line. The locus is a half line of points, with the origin as the starting point. $-2 - 2i$ has argument $-\dfrac{3\pi}{4}$ and so it is not on this half line.

(ii) Calculating $z - 2$ for each point and finding the argument of $(z - 2)$ gives:

(a) $z = 4 \qquad z - 2 = 2 \qquad \arg(z - 2) = 0$

(b) $z = 3 + i \qquad z - 2 = 1 + i \qquad \arg(1 + i) = \dfrac{\pi}{4}$

(c) $z = 4i \qquad z - 2 = 4i - 2 \quad \arg(-2 + 4i) = 2.03$

(d) $z = 8 + 6i \quad z - 2 = 6 + 6i \quad \arg(6 + 6i) = \dfrac{\pi}{4}$

(e) $z = 1 - i \qquad z - 2 = -1 - i \quad \arg(-1 - i) = -\dfrac{\pi}{4}$

So $\arg(z - 2) = \dfrac{\pi}{4}$ is satisfied by $z = 3 + i$ and $z = 8 + 6i$.

(iii) $z - 2$ represents a line between the point z and the point with coordinates $(2, 0)$.

So $\arg(z - 2) = \dfrac{\pi}{4}$ represents a line of points from $(2, 0)$ with an argument of $\dfrac{\pi}{4}$. This is a half line of points as shown.

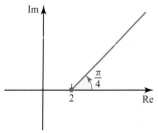

The line is a half line because points on the other half of the line would have an argument of $-\dfrac{\pi}{4}$ as was the case in part (ii)(e).

Activity 6.5 (Page 114)

The condition can be written as
$|z - (3 + 4i)| = |z - (-1 + 2i)|$.
$|z - (3 + 4i)|$ is the distance of point z from the point
$3 + 4i$ (point A) and $|z - (-1 + 2i)|$ is the distance of
point z from the point $-1 + 2i$ (point B).

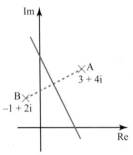

These distances are equal if z is on the
perpendicular bisector of AB.

Exercise 6.3 (Page 117)

1 (i)

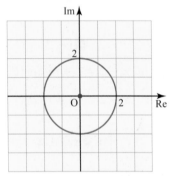

(ii)

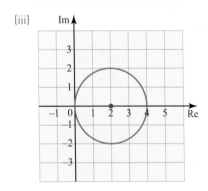

(iii)

(iv)

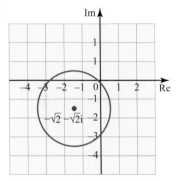

2 (i)

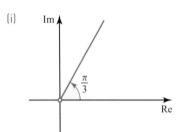

(ii)

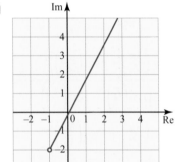

(iii)

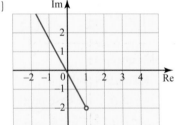

(iv)

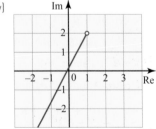

3 (i)

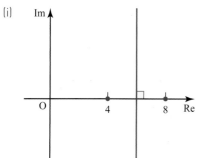

(ii)

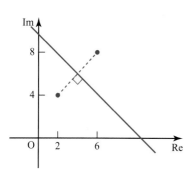

(iii)

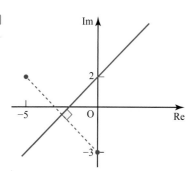

(iv)

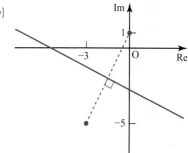

4 (i) $\left|z - (1 + i)\right| = 3$

(ii) $\arg(z + 2i) = \dfrac{3\pi}{4}$

(iii) $\left|z + 1\right| = \left|z - (3 + 2i)\right|$

5 (i) $\left|z - (4 + i)\right| \leqslant \left|z - (1 + 6i)\right|$

(ii) $-\dfrac{\pi}{4} \leqslant \arg(z + 2 - i) < 0$

(iii) $\left|z - (-2 + 3i)\right| < 4$

6

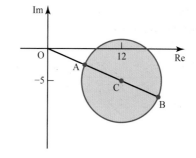

$|z|$ is least at A and greatest at B. Using Pythagoras' theorem, the distance OC is $\sqrt{(-5)^2 + 12^2} = 13$. We know AC = 7 and so OA = 13 − 7 = 6.

So, minimum value of $|z|$ is OA = 6 and maximum value of $|z|$ is OB = 6 + 14 = 20.

7 (i)

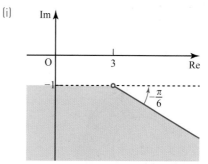

(ii)

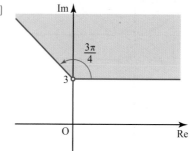

(iii)

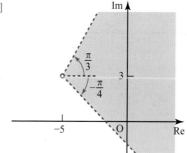

8 (i)

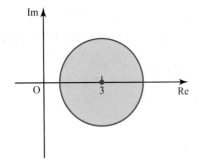

(ii)

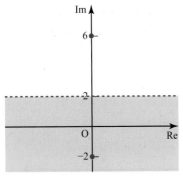

(iii)

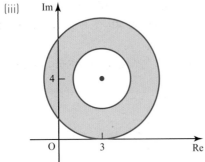

(iv)

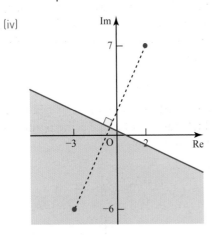

9

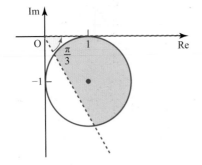

10 (i) (a) $|z + 1 + 2i| = 3$

(b) $|z + 6| = |z + 4i|$

(ii) $|z + 1 + 2i| \leq 3$ and $|z + 6| \geq |z + 4i|$

11

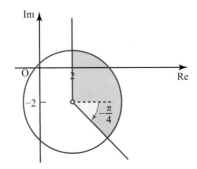

12 (i) $\dfrac{2\pi}{3}, 2$

(ii)

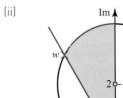

13

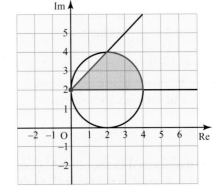

14

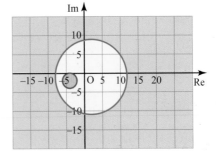

There are no values of z which satisfy both regions simultaneously.

15 The diagram shows $|z - 5 + 4i| = 3$. The minimum value is 7 and the maximum value is 13.

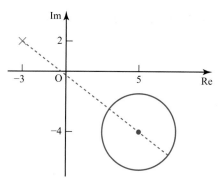

16 (i) Centre is $(3, -12)$, radius 6

(ii)
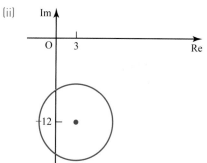

Chapter 7

Discussion point (Page 122)

The triangles are all similar to each other.
256 yellow triangles make up the blue triangle.

Activity 7.1 (Page 123)

The diagram shows the image of the unit square OABC under the transformation with matrix $\begin{pmatrix} a & b \\ c & d \end{pmatrix}$.

The point A$(1, 0)$ is transformed to the point A′ (a, c); the point C$(0, 1)$ is transformed to the point C′ (b, d). B′ has coordinates $(a + b, c + d)$.

The area of the parallelogram is given by the area of the whole rectangle minus the area of the rectangles and triangles.

Area of rectangle $= b \times c$

Area of first triangle $= \dfrac{1}{2} \times b \times d$

Area of second triangle $= \dfrac{1}{2} \times a \times c$

Area of whole rectangle $= (a + b) \times (c + d)$

Therefore the area of the parallelogram is

$(a + b) \times (c + d) - 2\left(bc + \dfrac{1}{2}bd + \dfrac{1}{2}ac\right) = ad - bc$.

Discussion point (Page 125)

(i) A rotation does not reverse the order of the vertices, e.g. for $\mathbf{A} = \begin{pmatrix} 0 & -1 \\ 1 & 0 \end{pmatrix}$,

det $\mathbf{A} = 1$ which is positive.

(ii) A reflection reverses the order of the vertices, e.g. for $\mathbf{B} = \begin{pmatrix} 1 & 0 \\ 0 & -1 \end{pmatrix}$,

det $\mathbf{B} = -1$ which is negative.

(iii) An enlargement does not reverse the order of the vertices, e.g. for $\mathbf{C} = \begin{pmatrix} 2 & 0 \\ 0 & 2 \end{pmatrix}$,

det $\mathbf{C} = 4$ which is positive

Discussion point (Page 126)

The matrix has determinant 8.

8 represents the volume scale factor of the transformation in three dimensions. If you think about the effect of the transformation represented by the matrix $\begin{pmatrix} 2 & 0 & 0 \\ 0 & 2 & 0 \\ 0 & 0 & 2 \end{pmatrix}$ on each of the unit vectors

$\mathbf{i} = \begin{pmatrix} 1 \\ 0 \\ 0 \end{pmatrix}$, $\mathbf{j} = \begin{pmatrix} 0 \\ 1 \\ 0 \end{pmatrix}$ and $\mathbf{k} = \begin{pmatrix} 0 \\ 0 \\ 1 \end{pmatrix}$,

the three edges of the unit cube have each increased by a length scale factor of 2. The overall effect would be that the volume would increase by a scale factor of $2 \times 2 \times 2 = 8$.

Discussion point (Page 127)

A 3×3 matrix with zero determinant will produce an image which has no volume, i.e. the points are all mapped to the same plane.

Exercise 7.1 (Page 127)

1 (i) (a)

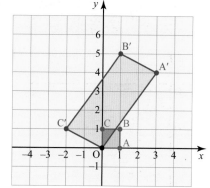

(b) Area of parallelogram = 11
(c) 11

(ii) (a)

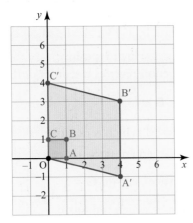

(b) Area of parallelogram = 16
(c) 16

(iii) (a)

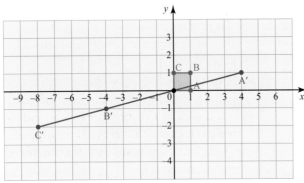

(b) area of parallelogram = 0
(c) 0

(iv) (a)

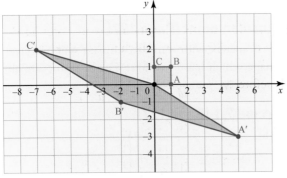

(b) area of parallelogram = 11
(c) −11

2 $x = 2, x = 6$

3 (i) $\mathbf{A} = \begin{pmatrix} 1 & 0 \\ 0 & -1 \end{pmatrix}$ $\mathbf{C} = \begin{pmatrix} 0 & 1 \\ 1 & 0 \end{pmatrix}$

$\mathbf{B} = \begin{pmatrix} -1 & 0 \\ 0 & 1 \end{pmatrix}$ $\mathbf{D} = \begin{pmatrix} 0 & -1 \\ -1 & 0 \end{pmatrix}$

(ii) no solution required

(iii) A

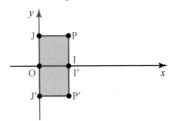

B

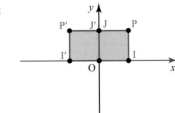

C

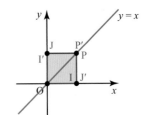

D

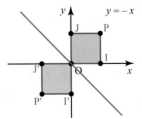

4 66 cm²

5 $ad = 1$

6 (i) $\begin{pmatrix} 1 & 3 \\ 0 & 1 \end{pmatrix}$

(ii) determinant = 1 so area is preserved

7 determinant = 6 so volume of image
$6 \times 5 = 30 \, \text{cm}^3$

8 (i) det $\mathbf{M} = -2$, det $\mathbf{N} = 7$

(ii) $\mathbf{MN} = \begin{pmatrix} 9 & 13 \\ 8 & 10 \end{pmatrix}$, $\qquad$ det $(\mathbf{MN}) = -14$

and $-14 = -2 \times 7$

9 (i)

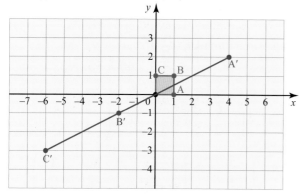

(ii) The image of all points lie on the line. $y = \frac{1}{2}x$

The determinant of the matrix is zero which shows that the image will have zero area.

10 (i) $\begin{pmatrix} 5p - 10q \\ -p + 2q \end{pmatrix}$

(ii) $y = -\frac{1}{5}x$

(iii) det $\mathbf{N} = 0$ and so the image has zero area

11 (i) $\mathbf{T} = \begin{pmatrix} 1 & 2 \\ 3 & 6 \end{pmatrix}$,

det $\mathbf{T} = (1 \times 6) - (3 \times 2) = 0$

(ii) $\begin{pmatrix} x' \\ y' \end{pmatrix} = \begin{pmatrix} 1 & 2 \\ 3 & 6 \end{pmatrix}\begin{pmatrix} x \\ y \end{pmatrix}$

$\Rightarrow \begin{matrix} x' = x + 2y \\ y' = 3x + 6y \end{matrix} \Rightarrow y' = 3x'$

(iii) $(3, 9)$

12 $\begin{pmatrix} x' \\ y' \end{pmatrix} = \begin{pmatrix} a & b \\ c & d \end{pmatrix}\begin{pmatrix} x \\ y \end{pmatrix} \Rightarrow \begin{pmatrix} x' \\ y' \end{pmatrix}$

$= \begin{pmatrix} ax + by \\ cx + dy \end{pmatrix} \Rightarrow \begin{matrix} x' = ax + by \\ y' = cx + dy \end{matrix}$

Solving simultaneously and using the fact $ad - bc = 0$ gives the result.

13 (i) $y = 3x - 3s + t$

(ii) $\mathrm{P}'\left(\frac{9}{8}s - \frac{3}{8}t, \frac{3}{8}s - \frac{1}{8}t\right)$

(iii) $\begin{pmatrix} \frac{9}{8} & -\frac{3}{8} \\ \frac{3}{8} & -\frac{1}{8} \end{pmatrix}$ which has determinant

$\left(\frac{9}{8} \times -\frac{1}{8}\right) - \left(\frac{3}{8} \times -\frac{3}{8}\right) = 0$

Activity 7.2 (Page 126)

$\det(\mathbf{A}) = (1 \times 3) - (-2 \times 4) = 11$

$\det(\mathbf{B}) = (3 \times 2) - (2 \times 1) = 4$

$\mathbf{AB} = \begin{pmatrix} 1 & 4 \\ -2 & 3 \end{pmatrix}\begin{pmatrix} 3 & 1 \\ 2 & 2 \end{pmatrix} = \begin{pmatrix} 11 & 9 \\ 0 & 4 \end{pmatrix}$

$\det(\mathbf{AB}) = 44 = 11 \times 4 = \det(\mathbf{A}) \times \det(\mathbf{B})$

Activity 7.3 (Page 130)

(i) $\mathbf{P} = \begin{pmatrix} 1 & 0 \\ 0 & -1 \end{pmatrix}$

(ii) $\mathbf{P}^2 = \begin{pmatrix} 1 & 0 \\ 0 & 1 \end{pmatrix}$

(iii) Reflecting an object in the x-axis twice takes it back to the starting position and so the final image is the same as the original object. Hence the matrix for the combined transformation is $\mathbf{I}$.

Activity 7.4 (Page 131)

$\begin{pmatrix} a & b \\ c & d \end{pmatrix}\begin{pmatrix} d & -b \\ -c & a \end{pmatrix} = \begin{pmatrix} ad - bc & 0 \\ 0 & ad - bc \end{pmatrix}$

To turn this into the identity matrix it would need to be divided by $ad - bc$ which is the value $|\mathbf{M}|$.

Therefore $\mathbf{M}^{-1} = \dfrac{1}{ad - bc}\begin{pmatrix} d & -b \\ -c & a \end{pmatrix}$

Activity 7.5 (Page 132)

(i) $\mathbf{AA}^{-1} = \frac{1}{4}\begin{pmatrix} 11 & 3 \\ 6 & 2 \end{pmatrix}\begin{pmatrix} 2 & -3 \\ -6 & 11 \end{pmatrix}$

$= \frac{1}{4}\begin{pmatrix} 4 & 0 \\ 0 & 4 \end{pmatrix} = \mathbf{I}$

$\mathbf{A}^{-1}\mathbf{A} = \frac{1}{4}\begin{pmatrix} 2 & -3 \\ -6 & 11 \end{pmatrix}\begin{pmatrix} 11 & 3 \\ 6 & 2 \end{pmatrix}$

$= \frac{1}{4}\begin{pmatrix} 4 & 0 \\ 0 & 4 \end{pmatrix} = \mathbf{I}$

(ii) $\mathbf{M}^{-1} = \dfrac{1}{ad - bc}\begin{pmatrix} d & -b \\ -c & a \end{pmatrix}$

$\mathbf{MM}^{-1} = \begin{pmatrix} a & b \\ c & d \end{pmatrix} \dfrac{1}{ad - bc}\begin{pmatrix} d & -b \\ -c & a \end{pmatrix}$

$= \dfrac{1}{ad - bc}\begin{pmatrix} a & b \\ c & d \end{pmatrix}\begin{pmatrix} d & -b \\ -c & a \end{pmatrix}$

$= \dfrac{1}{ad - bc}\begin{pmatrix} ad - bc & -ab + ab \\ cd - dc & -cb + ad \end{pmatrix}$

$= \dfrac{1}{ad - bc}\begin{pmatrix} ad - bc & 0 \\ 0 & ad - bc \end{pmatrix} = \begin{pmatrix} 1 & 0 \\ 0 & 1 \end{pmatrix} = \mathbf{I}$

$\mathbf{M}^{-1}\mathbf{M} = \dfrac{1}{ad - bc}\begin{pmatrix} d & -b \\ -c & a \end{pmatrix}\begin{pmatrix} a & b \\ c & d \end{pmatrix}$

$= \dfrac{1}{ad - bc}\begin{pmatrix} da - bc & db - bd \\ -ca + ac & -cb + ad \end{pmatrix}$

$= \dfrac{1}{ad - bc}\begin{pmatrix} ad - bc & 0 \\ 0 & ad - bc \end{pmatrix}$

$= \begin{pmatrix} 1 & 0 \\ 0 & 1 \end{pmatrix} = \mathbf{I}$

Discussion point (Page 132)

First reverse the reflection by using the transformation with the inverse matrix of the reflection. Secondly, reverse the rotation by using the transformation with the inverse matrix of the rotation.

$(\mathbf{MN})^{-1} = \mathbf{N}^{-1}\mathbf{M}^{-1}$

Exercise 7.2 (Page 133)

1　(i)　$(10, -6)$

(ii)　$-\dfrac{1}{2}\begin{pmatrix} 0 & 1 \\ 2 & 5 \end{pmatrix}$

(iii)　$(1, 2)$

2　(i)　non-singular, $\dfrac{1}{24}\begin{pmatrix} 2 & -3 \\ 4 & 6 \end{pmatrix}$

(ii)　singular

(iii)　non-singular, $\dfrac{1}{112}\begin{pmatrix} 11 & -3 \\ -3 & 11 \end{pmatrix}$

(iv)　singular

(v)　singular

(vi)　singular

(vii)　non-singular, $\dfrac{1}{16(1 - ab)}\begin{pmatrix} -8 & -4a \\ -4b & -2 \end{pmatrix}$

3　(i)　non-singular, $\dfrac{1}{140}\begin{pmatrix} 21 & 10 & 27 \\ -7 & -50 & -9 \\ 14 & 20 & -2 \end{pmatrix}$

(ii)　singular

(iii)　non-singular, $\dfrac{1}{121}\begin{pmatrix} -17 & 15 & 6 \\ -91 & 2 & 25 \\ 46 & -5 & -2 \end{pmatrix}$

4　(i)　$\dfrac{1}{3}\begin{pmatrix} 3 & -6 \\ -2 & 5 \end{pmatrix}$　(ii)　$\dfrac{1}{2}\begin{pmatrix} -1 & -5 \\ 2 & 8 \end{pmatrix}$

(iii)　$\begin{pmatrix} 28 & 19 \\ 10 & 7 \end{pmatrix}$　(iv)　$\begin{pmatrix} 50 & 63 \\ -12 & -15 \end{pmatrix}$

(v)　$\dfrac{1}{6}\begin{pmatrix} 7 & -19 \\ -10 & 28 \end{pmatrix}$　(vi)　$\dfrac{1}{6}\begin{pmatrix} -15 & -63 \\ 12 & 50 \end{pmatrix}$

(vii)　$\dfrac{1}{6}\begin{pmatrix} -15 & -63 \\ 12 & 50 \end{pmatrix}$　(viii)　$\dfrac{1}{6}\begin{pmatrix} 7 & -19 \\ -10 & 28 \end{pmatrix}$

5　(i)　$\dfrac{1}{8}\begin{pmatrix} 2 & 2 \\ -3 & 1 \end{pmatrix}$

(ii)　$\mathbf{M}^{-1}\mathbf{M} = \dfrac{1}{8}\begin{pmatrix} 2 & 2 \\ -3 & 1 \end{pmatrix}\begin{pmatrix} 1 & -2 \\ 3 & 2 \end{pmatrix} = \mathbf{I}$

6　$k = 2$ or $k = 3$

7　$\begin{pmatrix} 2 & 1 & 0 & -1 \\ 1 & 0 & -3 & 4 \end{pmatrix}$

8　(i)　$(3, 1), (1, 1)$ and $(-6, -2)$

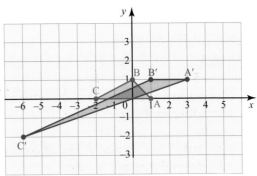

(ii)　ratio of area T' to T is $3 : 1.5$ or $2 : 1$

This is equal to the determinant of the matrix $\mathbf{M}$.

(iii)　$\mathbf{M}^{-1} = \dfrac{1}{2}\begin{pmatrix} 1 & -1 \\ -1 & 3 \end{pmatrix}$

9 (i) no solution required

(ii) $\mathbf{M}^n = (a + d)^{n-1}\mathbf{M}$

10 (i) No solution required

(ii) No solution required

(iii) $\begin{pmatrix} 0 & \dfrac{1}{5} \\ \dfrac{1}{6} & -\dfrac{11}{10} \end{pmatrix}$ (iv) $\begin{pmatrix} 33 & 6 \\ 5 & 0 \end{pmatrix}$

11 (i) $\begin{pmatrix} 18 & -9 & 4 \\ 1 & -7 & 2 \\ -1 & -4 & 1 \end{pmatrix}$

(ii) $\begin{pmatrix} 1 & a+7 & b+7c+4 \\ 0 & 1 & c+2 \\ 0 & 0 & 1 \end{pmatrix}$

(iii) $\begin{pmatrix} 1 & -7 & 10 \\ 0 & 1 & -2 \\ 0 & 0 & 1 \end{pmatrix}$

(iv) $\begin{pmatrix} 1 & 0 & 0 \\ -3 & 1 & 0 \\ -11 & 4 & 1 \end{pmatrix}$

(v) $\begin{pmatrix} 1 & -7 & 10 \\ -3 & 22 & -32 \\ -11 & 81 & -117 \end{pmatrix}$

12 $k = 3$

Exercise 7.3 (Page 139)

1 (i) (a) 5   (b) 5

(ii) (a) −5   (b) −5

Interchanging the rows and columns has not changed the determinant.

(iii) (a) 0   (b) 0

If a matrix has a repeated row or column the determinant will be zero.

2 (i) $\dfrac{1}{3}\begin{pmatrix} 3 & 0 & -6 \\ -4 & 2 & 3 \\ 2 & -1 & 0 \end{pmatrix}$

(ii) matrix is singular

(iii) $\begin{pmatrix} -0.06 & -0.1 & -0.1 \\ 0.92 & 0.2 & 0.7 \\ 0.66 & 0.1 & 0.6 \end{pmatrix}$

(iv) $\dfrac{1}{21}\begin{pmatrix} 34 & 11 & 32 \\ 9 & 6 & 6 \\ -38 & -16 & -37 \end{pmatrix}$

3 $\mathbf{M}^{-1} = \dfrac{1}{28-10k}\begin{pmatrix} 4 & -10 & 12 \\ -(4k-8) & 8 & -(4+2k) \\ -k & 7 & -3k \end{pmatrix}$;

$k = 2.8$

4 (i) The columns of the matrix have been moved one place to the right, with the final column moving to replace the first. This is called **cyclical interchange** of the columns.

(ii) det $\mathbf{A}$ = det $\mathbf{B}$ = det $\mathbf{C}$ = −26
Cyclical interchange of the columns leaves the determinant unchanged.

5 $x = \dfrac{-1 \pm \sqrt{41}}{2}$

6 $x = 1, x = 4$

7 $1 < k < 5$

8 (i) Let $\mathbf{X} = (\mathbf{PQ})^{-1}$ so $\mathbf{X}(\mathbf{PQ}) = \mathbf{I}$.

$\Rightarrow \mathbf{X}(\mathbf{PQ})\mathbf{Q}^{-1} = \mathbf{IQ}^{-1} = \mathbf{Q}^{-1}$

$\Rightarrow \mathbf{XP}(\mathbf{QQ}^{-1}) = \mathbf{XP} = \mathbf{Q}^{-1}$

$\Rightarrow \mathbf{XPP}^{-1} = \mathbf{Q}^{-1}\mathbf{P}^{-1}$

$\Rightarrow \mathbf{X} = \mathbf{Q}^{-1}\mathbf{P}^{-1}$

(ii) $\mathbf{P}^{-1} = \begin{pmatrix} -\dfrac{1}{9} & \dfrac{1}{6} & -\dfrac{4}{9} \\ \dfrac{2}{9} & \dfrac{1}{6} & -\dfrac{1}{9} \\ -\dfrac{1}{3} & \dfrac{1}{2} & -\dfrac{1}{3} \end{pmatrix}$

$\mathbf{Q}^{-1} = \begin{pmatrix} \dfrac{3}{2} & -4 & \dfrac{1}{2} \\ 1 & -2 & 0 \\ -\dfrac{3}{2} & 5 & -\dfrac{1}{2} \end{pmatrix}$

$(\mathbf{PQ})^{-1} = \mathbf{Q}^{-1}\mathbf{P}^{-1} = \begin{pmatrix} -\dfrac{11}{9} & -\dfrac{1}{6} & -\dfrac{7}{18} \\ -\dfrac{5}{9} & -\dfrac{1}{6} & -\dfrac{2}{9} \\ \dfrac{13}{9} & \dfrac{1}{3} & \dfrac{5}{18} \end{pmatrix}$

9 (i) no solution required

(ii) Multiplying only the first column by k equates to a stretch of scale factor k in one direction, so only multiplies the volume by k.

(iii) Multiplying any column by k multiplies the determinant by k.

10 (i) $10 \times 43 = 430$

(ii) $4 \times 5 \times -7 \times 43 = -6020$

(iii) $x \times 2 \times y \times 43 = 86xy$

(iv) $x^4 \times \dfrac{1}{2x} \times 4y \times 43 = 86x^3y$

Activity 7.6 (Page 143)

(i)
$$\begin{pmatrix} 2 & -2 & 3 \\ 5 & 1 & -1 \\ 3 & 4 & -2 \end{pmatrix} \begin{pmatrix} x \\ y \\ z \end{pmatrix} = \begin{pmatrix} 4 \\ -6 \\ 1 \end{pmatrix}$$

$$\begin{pmatrix} x \\ y \\ z \end{pmatrix} = \begin{pmatrix} 2 & -2 & 3 \\ 5 & 1 & -1 \\ 3 & 4 & -2 \end{pmatrix}^{-1} \begin{pmatrix} 4 \\ -6 \\ 1 \end{pmatrix}$$

$$\Rightarrow x = -1, \ y = 3, \ z = 4$$

(ii)
$$\begin{pmatrix} 2 & -2 & 3 \\ 5 & 1 & -1 \\ 3 & 3 & -4 \end{pmatrix} \begin{pmatrix} x \\ y \\ z \end{pmatrix} = \begin{pmatrix} 4 \\ -6 \\ 1 \end{pmatrix}$$

$$\begin{pmatrix} x \\ y \\ z \end{pmatrix} = \begin{pmatrix} 2 & -2 & 3 \\ 5 & 1 & -1 \\ 3 & 3 & -4 \end{pmatrix}^{-1} \begin{pmatrix} 4 \\ -6 \\ 1 \end{pmatrix}$$

The equations cannot be solved as the determinant of the matrix is zero, so the inverse matrix does not exist. Using an algebraic method results in inconsistent equations, which have no solutions.

Exercise 7.4 (Page 144)

1 (i) $\dfrac{1}{11}\begin{pmatrix} 3 & 1 \\ -2 & 3 \end{pmatrix}$

(ii) $x = 1, \ y = 1$

2 (i) $x = 2, \ y = -1$

(ii) $x = 4, \ y = 1.5$

3 (i) $\begin{pmatrix} 0.5 & 0 & -0.5 \\ -0.8 & -0.2 & 1.4 \\ 0.3 & 0.2 & 0.1 \end{pmatrix}$

(ii) $x = -2, \ y = 4.6, \ z = -0.6$

4 $x = 4, \ y = -1, \ z = 1$

5 (i) Single point of intersection at $(8.5, -1.5)$

(ii) Lines are coincident. There are an infinite number of solutions of the form $(6 - 2\lambda, \ \lambda)$.

(iii) Lines are parallel and therefore there are no solutions.

6 $k = 4$, infinite number of solutions

$k = -4$, no solutions

7 (i) $\begin{pmatrix} k-1 & 0 & 0 \\ 0 & k-1 & 0 \\ 0 & 0 & k-1 \end{pmatrix}$ so

$$\mathbf{A}^{-1} = \dfrac{1}{k-1}\begin{pmatrix} -1 & 3k+8 & 4k+10 \\ -2 & 2k+20 & 3k+25 \\ 1 & -11 & -14 \end{pmatrix}$$

where $k \neq 1$

(ii) $x = 8, \ y = 6, \ z = 0$

8 $a \neq \pm b$

$$x = \dfrac{3 + b^2}{b^2 - 9}, \ y = \dfrac{4b}{b^2 - 9}$$

$b = -1$ or -3 but since $a \neq \pm b$, $b = -1$

Chapter 8

Discussion point (Page 147)

The architecture of the Eden project is described in http://www.edenproject.com/eden-story/behind-the-scenes/architecture-at-eden.

Discussion point (Page 148)

Could also draw lines perpendicular to the x-axis to the points P and Q as shown.

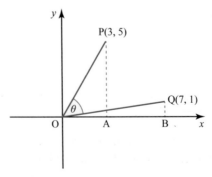

Using trigonometry on the two right-angled triangles, find $\angle POA$ and $\angle QOB$ and calculate the difference between these values, which equals θ.

Activity 8.1 (Page 149)

$$\cos\theta = \frac{\left|\overrightarrow{OA}\right|^2 + \left|\overrightarrow{OB}\right|^2 - \left|\overrightarrow{AB}\right|^2}{2 \times \left|\overrightarrow{OA}\right| \times \left|\overrightarrow{OB}\right|}$$

$$\Rightarrow \cos\theta = \frac{\left(a_1^2 + a_2^2\right) + \left(b_1^2 + b_2^2\right) - \left[\left(b_1 - a_1\right)^2\right] + \left[\left(b_2 - a_2\right)^2\right]}{2\left(a_1^2 + a_2^2\right)\left(b_1^2 + b_2^2\right)}$$

$$\Rightarrow \cos\theta = \frac{2\left(a_1 b_1 + a_2 b_2\right)}{2\sqrt{\left(a_1^2 + a_2^2\right)}\sqrt{\left(b_1^2 + b_2^2\right)}} = \frac{a_1 b_1 + a_2 b_2}{|\mathbf{a}||\mathbf{b}|}$$

Discussion point (Page 151)

$$\overrightarrow{BA} = \begin{pmatrix} 2 \\ -2 \\ -4 \end{pmatrix} \qquad \overrightarrow{BC} = \begin{pmatrix} 8 \\ -7 \\ 0 \end{pmatrix}$$

$$\begin{pmatrix} 2 \\ -2 \\ -4 \end{pmatrix} \cdot \begin{pmatrix} 8 \\ -7 \\ 0 \end{pmatrix} = 30 \text{ which is the same answer as in Example 7.2}$$

Exercise 8.1 (Page 152)

1. (i) -4 (ii) 4
   (iii) 1 (iv) 7
2. $64.7°$
3. (i) $66.6°$
   (ii) $113.4°$
   (iii) $113.4°$
4. (i) $0°$ The vectors are parallel.
   (ii) $180°$ The vectors are in opposite directions (one is a negative multiple of the other).
5. -17
6. $-2, -3$
7. $52.2°, 33.2°, 94.6°$
8. $35.8°, 71.1°, 60.9°$
9. (i) $(0, 4, 3)$
   (ii) $\begin{pmatrix} -5 \\ 4 \\ 3 \end{pmatrix}, 5\sqrt{2}$
   (iii) $25.1°$
10. (i) $A(4, 0, 0)$ $C(0, 5, 0)$
    $F(4, 0, 3)$ $H(0, 5, 3)$
    (ii) EPF not vertical as the points do not have the same y-coordinate. The roof sections form trapezia.
    (iii) $\cos\theta = -\frac{1}{3}$ Area $= 2\sqrt{2}$
    (iv) $68.9°$

12. $16 + (\mathbf{a} - \mathbf{b}) \cdot \mathbf{c}$

Activity 8.2 (Page 155)

(ii) $\begin{pmatrix} -2 \\ -9 \end{pmatrix}, \begin{pmatrix} 0 \\ -5 \end{pmatrix}, \begin{pmatrix} 2 \\ -1 \end{pmatrix}, \begin{pmatrix} 3 \\ 1 \end{pmatrix}, \begin{pmatrix} 3.5 \\ 2 \end{pmatrix}, \begin{pmatrix} 6 \\ 7 \end{pmatrix}, \begin{pmatrix} 8 \\ 11 \end{pmatrix}$

(iii) The points join to form the straight line $y = 2x - 5$

(iv) (a) It lies between the point A$(2, -1)$ and the point B$(4, 3)$
(b) It lies beyond the point B$(4, 3)$
(c) It lies beyond A in the opposite direction to the point B.

Activity 8.3 (Page 156)

(i) $x = a_1 + \lambda d_1$ $y = a_2 + \lambda d_2$

(ii) $\lambda = \dfrac{x - a_1}{d_1}$ $\lambda = \dfrac{y - a_2}{d_2}$

$$y = \frac{d_2}{d_1}x + \frac{a_2 d_1 - a_1 d_2}{d_1} \text{ so } m = \frac{d_2}{d_1} \text{ and}$$

$$c = \frac{a_2 d_1 - a_1 d_2}{d_1}$$

Activity 8.4 (Page 162)

$y = \dfrac{1}{2}x + 3$ and $y = -3x + 8$. You can find the angle between them by finding the difference between arctan(gradient) for each line.

Exercise 8.2 (Page 163)

1. (i) $\mathbf{r} = \begin{pmatrix} 3 \\ 1 \end{pmatrix} + \lambda \begin{pmatrix} 5 \\ -2 \end{pmatrix}$

   (ii) $\mathbf{r} = \begin{pmatrix} 5 \\ -1 \end{pmatrix} + \lambda \begin{pmatrix} 0 \\ 4 \end{pmatrix}$

   (iii) $\mathbf{r} = \begin{pmatrix} -2 \\ 4 \end{pmatrix} + \lambda \begin{pmatrix} 5 \\ 5 \end{pmatrix}$

   (iv) $\mathbf{r} = \begin{pmatrix} 0 \\ 8 \end{pmatrix} + \lambda \begin{pmatrix} -2 \\ -11 \end{pmatrix}$

2. (i) $\mathbf{r} = \begin{pmatrix} 2 \\ 4 \\ -1 \end{pmatrix} + \lambda \begin{pmatrix} 3 \\ 6 \\ 4 \end{pmatrix}$

   (ii) $\mathbf{r} = \begin{pmatrix} 1 \\ 0 \\ -1 \end{pmatrix} + \lambda \begin{pmatrix} 1 \\ 0 \\ 0 \end{pmatrix}$

(iii) $\mathbf{r} = \begin{pmatrix} 1 \\ 0 \\ 4 \end{pmatrix} + \lambda \begin{pmatrix} 5 \\ 3 \\ -6 \end{pmatrix}$

(iv) $\mathbf{r} = \begin{pmatrix} 0 \\ 0 \\ 1 \end{pmatrix} + \lambda \begin{pmatrix} 2 \\ 1 \\ 3 \end{pmatrix}$

3 (i) $\dfrac{x-2}{3} = \dfrac{y-4}{6} = \dfrac{z+1}{4}$

(ii) $x - 1 = \dfrac{y}{3} = \dfrac{z+1}{4}$

(iii) $x - 3 = \dfrac{z-4}{2}$ and $y = 0$

(iv) $\dfrac{x}{2} = \dfrac{z-1}{4}$ and $y = 4$

4 (i) $\mathbf{r} = \begin{pmatrix} 3 \\ -2 \\ 1 \end{pmatrix} + \lambda \begin{pmatrix} 5 \\ 3 \\ 4 \end{pmatrix}$

(ii) $\mathbf{r} = \begin{pmatrix} 0 \\ 0 \\ -1 \end{pmatrix} + \lambda \begin{pmatrix} 1 \\ 2 \\ 3 \end{pmatrix}$

(iii) $\mathbf{r} = \lambda \begin{pmatrix} 1 \\ 1 \\ 1 \end{pmatrix}$

(iv) $\mathbf{r} = \begin{pmatrix} 2 \\ 0 \\ 0 \end{pmatrix} + \lambda \begin{pmatrix} 0 \\ 1 \\ 1 \end{pmatrix}$

5 $\mathbf{r} = \begin{pmatrix} 3 \\ -5 \\ 2 \end{pmatrix} + \lambda \begin{pmatrix} 0 \\ 1 \\ 0 \end{pmatrix}$;

$x = 3$, $z = 2$ and y can have any value

6 (i) $\begin{pmatrix} 4 \\ 1 \end{pmatrix}$ (ii) $\begin{pmatrix} 5 \\ 5 \end{pmatrix}$ (iii) $\begin{pmatrix} -5 \\ 6 \end{pmatrix}$

7 (i) Intersect at $(3, 2, -13)$
(ii) Lines are skew
(iii) Lines are parallel
(iv) Intersect at $(4, -7, 11)$
(v) Lines are skew

8 (i) $45°$ (ii) $56.3°$
(iii) $53.6°$ (iv) $81.8°$
(v) $8.7°$

9 Do not meet.
10 $6, 9, \sqrt{77}$

11 (i) $\begin{pmatrix} -0.25 \\ 0 \\ 0 \end{pmatrix}$

(ii) $C(0, 0.05, 1.1)$

(iii) DE: $\mathbf{r} = \begin{pmatrix} 0 \\ 0 \\ 1 \end{pmatrix} + \lambda \begin{pmatrix} 1 \\ 0 \\ 0 \end{pmatrix}$

EF: $\mathbf{r} = \begin{pmatrix} 0.25 \\ 0 \\ 1 \end{pmatrix} + \mu \begin{pmatrix} 0 \\ 1 \\ 2 \end{pmatrix}$

12 (i) $(1, 0.5, 0)$ (ii) $41.8°$
(iii) $027°$ (iv) $(2, 2.5, 2)$
(v) $t = 2$; $\sqrt{5}$ km

Discussion point (Page 166)

The pencil will be perpendicular to each line; it will make an angle of 90° with each line. If the table is tilted and the pencil is tilted with it, the pencil will still be perpendicular to each line, so it makes no difference.

Discussion point (Page 169)

One method would start by calculating the vectors $\overrightarrow{AB}$ and $\overrightarrow{AC}$. Use the scalar product to find a vector perpendicular to these two vectors, which can be used as the normal $\begin{pmatrix} n_1 \\ n_2 \\ n_3 \end{pmatrix}$ to the plane. Substitute one of the points A, B or C into the equation $n_1 x + n_2 y + n_3 z + d = 0$ to find the value of d.

Alternatively, substitute the three points into the equation $ax + by + cz + d = 0$ to form three simultaneous equations and use a matrix method to solve these equations and hence find the equation of the plane.

Activity 8.5 (Page 170)

$a + b + c = -d$

$a - b = -d$

$-a + 2c = -d$

$$\begin{pmatrix} 1 & 1 & 1 \\ 1 & -1 & 0 \\ -1 & 0 & 2 \end{pmatrix} \begin{pmatrix} a \\ b \\ c \end{pmatrix} = \begin{pmatrix} -d \\ -d \\ -d \end{pmatrix} \Rightarrow$$

$$\begin{pmatrix} a \\ b \\ c \end{pmatrix} = \begin{pmatrix} 0.4 & 0.4 & -0.2 \\ 0.4 & -0.6 & -0.2 \\ 0.2 & 0.2 & 0.4 \end{pmatrix} \begin{pmatrix} -d \\ -d \\ -d \end{pmatrix}$$

$$= \begin{pmatrix} -0.6d \\ 0.4d \\ -0.8d \end{pmatrix}$$

The plane has equation
$-0.6x + 0.4y - 0.8z + 1 = 0$.

Discussion point (Page 171)

The vectors **b** and **c** must not be multiples of one another. Provided they have different directions any choice of **b** and **c** can be used to describe the direction of a plane.

Exercise 8.3 (Page 173)

1 (i) $\begin{pmatrix} 5 \\ -3 \\ 2 \end{pmatrix}$

   (ii) $(5 \times 1) - (3 \times 4) + (2 \times 3) + 1 = 0$

2 (i) $\mathbf{r}.\begin{pmatrix} 1 \\ 1 \\ 1 \end{pmatrix} = 6$ (ii) $\mathbf{r}.\begin{pmatrix} 1 \\ 1 \\ 1 \end{pmatrix} = 0$

   (iii) $\mathbf{r}.\begin{pmatrix} -1 \\ -1 \\ -1 \end{pmatrix} = -6$ (iv) $\mathbf{r}.\begin{pmatrix} 2 \\ 2 \\ 2 \end{pmatrix} = 16$

3 (i) $x + y + z = 6$ (ii) $x + y + z = 0$

   (iii) $-x - y - z = -6$

   (iv) $2x + 2y + 2z = 16$

   The planes are parallel to each other; parts (i) and (iii) represent the same plane.

4 (i) $\mathbf{r} = \begin{pmatrix} 1 \\ 1 \\ 1 \end{pmatrix} + \lambda \begin{pmatrix} 1 \\ 3 \\ 4 \end{pmatrix} + \mu \begin{pmatrix} -2 \\ 1 \\ -4 \end{pmatrix}$

   or any equivalent form.

   (ii) Do not know the direction of the normal vector.

5 (i) 80.4°   (ii) 90°   (iii) 69.9°

6   $\mathbf{r}.(-\mathbf{i} + 3\mathbf{j} - 2\mathbf{k}) = 5$ $-x + 3y - 2z = 5$

7   $4x - 5y + 6z + 29 = 0$

8   $-1, 4$

9   $\dfrac{16 + 9\sqrt{6}}{5}$

10   $x - 4y + 7z = 27$

11   $\begin{pmatrix} -1 \\ 2 \\ 2 \end{pmatrix}.\begin{pmatrix} -2 \\ 4 \\ -5 \end{pmatrix} = 2 + 8 - 10 = 0$ and

   $\begin{pmatrix} 2 \\ 1 \\ 0 \end{pmatrix}.\begin{pmatrix} -2 \\ 4 \\ -5 \end{pmatrix} = -4 + 3 = 0$

   So $\begin{pmatrix} -2 \\ 4 \\ 5 \end{pmatrix}$ is perpendicular to both

   $\begin{pmatrix} -1 \\ 2 \\ 2 \end{pmatrix}$ and $\begin{pmatrix} 2 \\ 1 \\ 0 \end{pmatrix}$, which means that it is

   normal to both planes, so they have the same direction.

   The planes are distinct since the point $(2, 0, -1)$ lies in the first plane but $\begin{pmatrix} 2 \\ 0 \\ -1 \end{pmatrix}.\begin{pmatrix} -2 \\ 4 \\ 5 \end{pmatrix} =$

   $-9 \ (\neq 6)$ so the planes are parallel.

12 (i) $\overrightarrow{AB} = \begin{pmatrix} 2 \\ 2 \\ -2 \end{pmatrix}$ $\overrightarrow{AC} = \begin{pmatrix} 5 \\ 2 \\ -1 \end{pmatrix}$

   (iii) $x - 4y - 3z = -2$

13 (iii) B

15 (i) $\begin{pmatrix} 2 \\ -3 \\ 2 \end{pmatrix}, 10$

(ii) e.g. $\begin{pmatrix} -1 \\ 2 \\ -1 \end{pmatrix}$

(iii) e.g.
$\left(\mathbf{r} - \left(-\mathbf{i} + 2\mathbf{j} - \mathbf{k}\right)\right).\left(2\mathbf{i} - 3\mathbf{j} + 2\mathbf{k}\right) = 0$

16 π_1 and π_3 are parallel.

π_2 is perpendicular to both π_1 and π_3.

Discussion point (Page 176)

Some examples are:

- three parallel planes – bookshelves

- intersection at a unique point – three walls meeting in the corner of a room

- sheaf of planes – pages of a book

- triangular prism – the two sloping walls and the floor of a loft located within a roof space, or the sides and base of a triangular-shaped tent.

Exercise 8.4 (Page 180)

1 (i) $\begin{pmatrix} 0.15 & 0.25 & -0.05 \\ -0.05 & 0.25 & -0.65 \\ 0.25 & -0.25 & 0.25 \end{pmatrix}$

(ii) $(4, -18, 10)$

2 (i) $\begin{pmatrix} 1 & 0 & -1 \\ 4 & -1 & -5 \\ -4.5 & 1.5 & 5.5 \end{pmatrix}$ $(-1, -12, 15)$

3 (i) Intersect at $(1.8, 3, -3.1)$
(ii) Do not intersect at a unique point
(iii) Intersect at $(3, -14, 8)$
(iv) Do not intersect at a unique point

4 (i) $\begin{pmatrix} -1 & 1 & 1 \\ 2 & 1 & 1 \\ 1 & 1 & 1 \end{pmatrix}\begin{pmatrix} x \\ y \\ z \end{pmatrix} = \begin{pmatrix} -1 \\ 6 \\ 4 \end{pmatrix}$

None of the planes are parallel and det $\mathbf{M} = 0$ so the planes form either a sheaf or a prism of planes.

(ii) P lies on the second and third planes but not on the first; the planes form a prism.

(iii) Changing the first plane to be $-x + y + z = 0$ means P lies on this plane too and so they now form a sheaf.

5 $\begin{pmatrix} 1 & 2 & -1 \\ 2 & 4 & 1 \\ 3 & 6 & -3 \end{pmatrix}\begin{pmatrix} x \\ y \\ z \end{pmatrix} = \begin{pmatrix} 6 \\ 5 \\ 8 \end{pmatrix}$

The third row is a multiple of the first row, but not a multiple of the second row; the first and third planes are parallel and the second plane cuts through them to form two parallel straight lines.

6 $k = 3$, $m = -2$

7 (i) $\begin{pmatrix} 3 & 4 & 1 \\ 2 & -1 & -1 \\ 5 & 14 & 5 \end{pmatrix}\begin{pmatrix} x \\ y \\ z \end{pmatrix} = \begin{pmatrix} 5 \\ 4 \\ 7 \end{pmatrix}$

det $\mathbf{M} = 0$ and none of the planes are parallel to each other, so they form a prism or a sheaf.

(ii) P lies on all three planes, so the planes form a sheaf.

8 (i) The planes intersect in the unique point $\left(-2, 4\frac{1}{3}, \frac{1}{3}\right)$.

(ii) $k = 2$, $m = -1$, $n = -2$ or any multiple of these would make planes 1 and 2 parallel and cut by the third plane.

(iii) The first plane is coincident (the same as) the third plane, the second plane cuts through this plane. In part (ii) there were two parallel planes but they would not be coincident unless the values $k = \frac{4}{3}$, $m = -\frac{2}{3}$, $n = -\frac{4}{3}$ were chosen.

9 There are 8 possible arrangements:

- The planes intersect in a unique point.

- Two planes are parallel and are cut by the third plane to form two parallel lines.

- All three planes are parallel.

- The planes form a prism where each pair of planes meets in a straight line.

- The planes form a sheaf with all three intersecting in one straight line.

- Two planes are coincident and the third cuts through them.

- All three planes are coincident.

- Two planes are coincident and the third is parallel to these.

Exercise 8.5 (Page 184)

2 (i) $(0, 1, 3), 67.8°$
(ii) $(1, 1, 1), 3.01°$
(iii) $(8, 4, 2), 12.6°$
(iv) $(0, 0, 0), 70.5°$

3 (i) $\mathbf{r} = \begin{pmatrix} 4 \\ 1 \\ 3 \end{pmatrix} + \lambda \begin{pmatrix} 2 \\ 3 \\ 5 \end{pmatrix}$ or

$\mathbf{r} = \begin{pmatrix} 6 \\ 4 \\ 8 \end{pmatrix} + \mu \begin{pmatrix} 2 \\ 3 \\ 5 \end{pmatrix}$ or equivalent

(ii) Intersect at $(0, -5, -7)$
(iii) $11.5°$

4 (i) $\mathbf{r} = \begin{pmatrix} 13 \\ 5 \\ 0 \end{pmatrix} + \lambda \begin{pmatrix} 3 \\ 1 \\ -2 \end{pmatrix}$

(ii) $(4, 2, 6)$
(iii) $\sqrt{126}$

5 (i) $\overrightarrow{AB} = \begin{pmatrix} 2 \\ -3 \\ 2 \end{pmatrix}$

$\overrightarrow{AC} = \begin{pmatrix} -5 \\ 2 \\ -1 \end{pmatrix}$ $x + 8y + 11z = 15$

(ii) $P\left(\dfrac{24}{13}, -1, \dfrac{25}{13}\right)$ $Q\left(\dfrac{267}{40}, \dfrac{-67}{40}, \dfrac{79}{40}\right)$
(iii) $R(6, -1, 4)$ (iv) $27.06°$ (v) 5.15

6 (i) $\overrightarrow{AA'} = \begin{pmatrix} 1 \\ 2 \\ -3 \end{pmatrix}$ so AA′ is perpendicular to

$x + 2y - 3z = 0$

M has coordinates $(1.5, 3, 2.5)$ and $1.5 + (2 \times 3) - (3 \times 2.5) = 0$ so lies in the plane.

(ii) $(0, 3, 2)$; one possible equation is

$\mathbf{r} = \begin{pmatrix} 2 \\ 4 \\ 1 \end{pmatrix} + \lambda \begin{pmatrix} -2 \\ -1 \\ 1 \end{pmatrix}$

(iii) $80.4°$

7 (i) $AB = \sqrt{29}$ $AC = 5$; $56.1°$; 11.2
(ii) (b) $4x - 3y + 10z = -12$

(iii) $\mathbf{r} = \begin{pmatrix} 0 \\ 4 \\ 5 \end{pmatrix} + \lambda \begin{pmatrix} 4 \\ -3 \\ 10 \end{pmatrix}$;

intersect at $(-1.6, 5.2, 1)$
(iv) 16.7

Activity 8.7 (Page 190)

Because it is perpendicular to both l_1 and l_2.

Exercise 8.6 (Page 192)

1 (i) $\sqrt{29}$ (ii) 7 (iii) $2\sqrt{26}$
2 (i) 13 (ii) $3\sqrt{5}$ (iii) 1
3 (i) 5 (ii) 5 (iii) $5\sqrt{2}$

4 (i) $\mathbf{r} = \begin{pmatrix} 3 \\ 1 \\ 0 \end{pmatrix} + \lambda \begin{pmatrix} 1 \\ -2 \\ -1 \end{pmatrix}$ (ii) $\sqrt{\dfrac{7}{3}}$

5 (ii) $\dfrac{1}{2}\sqrt{38}$

6 (i) 4, the lines are skew
(ii) 0.4, the lines are skew
(iii) 0, the lines intersect
(iv) $\sqrt{\dfrac{77}{6}}$, the lines are parallel

7 (i) $\sqrt{570}$ (ii) $M = (6, -16, -9)$

8 (i) $\dfrac{2\sqrt{5}}{15}$

(ii) $\mathbf{r} = \begin{pmatrix} 2 \\ 0 \\ -5 \end{pmatrix} + \lambda \begin{pmatrix} 4 \\ -5 \\ 2 \end{pmatrix}$

(iii) $M\left(\dfrac{82}{45}, \dfrac{10}{45}, -\dfrac{229}{45}\right)$

9 (i) AB: $\mathbf{r} = \begin{pmatrix} 2 \\ -3 \\ 4 \end{pmatrix} + \lambda \begin{pmatrix} -1 \\ 0 \\ 1 \end{pmatrix}$ and

CD: $\mathbf{r} = \begin{pmatrix} 0 \\ 3 \\ -2 \end{pmatrix} + \lambda \begin{pmatrix} 2 \\ 0 \\ 7 \end{pmatrix}$ or

equivalent; 6
(ii) $3.08\,\text{m}$ – No, the cable is not long enough

10 (ii) $7 + \dfrac{2}{3}k$

(iii) -10.5

(iv) $\mathbf{r} = \begin{pmatrix} 4 \\ 12 \\ 5 \end{pmatrix} + \alpha \begin{pmatrix} 2 \\ 10 \\ 11 \end{pmatrix}$ or equivalent

11 (i) (b) $\begin{pmatrix} 2 \\ 2 \\ -1 \end{pmatrix}$

(c) $\mathbf{r} = \begin{pmatrix} 0 \\ -2 \\ 0 \end{pmatrix} + \lambda \begin{pmatrix} 2 \\ 2 \\ -1 \end{pmatrix}$

(ii) $\dfrac{46}{5}$ (iii) 15

(iv) $k = 50$; $(16, 8, 4)$

Practice Questions 2 (Page 197)

1 (i) Reflection in the line $y = 0$/the x-axis. [1]

(ii) Reflection in the line $x = 0$/the y-axis. [1]

(iii) $\mathbf{BA} = \begin{pmatrix} -1 & 0 \\ 0 & -1 \end{pmatrix}$. [1]

It represents rotation of $180°$ about the origin. [1]

(iv) $(\mathbf{BA})^{-1} = \begin{pmatrix} -1 & 0 \\ 0 & -1 \end{pmatrix} = \mathbf{BA}$. [1], [1]

A rotation of 180° about the origin followed by another rotation of 180° about the origin. Is equivalent to one full turn about the origin, which has no effect. This means, the inverse of a rotation of 180° about the origin is another rotation of 180° about the origin. [1]

2 (i) $z_1 z_2 = (a + bi)(c + di)$ [1], [1]

$= ac - bd + (ad + bc)i$

(ii) $|z_1| = \sqrt{a^2 + b^2}$, $|z_1| = \sqrt{c^2 + d^2}$ [1]

(iii)

$|z_1 z_2| = \sqrt{(ac - bd)^2 + (ad + bc)^2}$ [1]

$= \sqrt{a^2 c^2 - 2abcd + b^2 d^2 + a^2 d^2 + 2abcd + b^2 c^2}$

$= \sqrt{a^2 c^2 + b^2 d^2 + a^2 d^2 + b^2 c^2}$ [1]

$|z_1||z_2| = \sqrt{a^2 + b^2}\sqrt{c^2 + d^2}$

$= \sqrt{(a^2 + b^2)(c^2 + d^2)}$

$= \sqrt{a^2 c^2 + b^2 d^2 + a^2 d^2 + b^2 c^2}$ [1]

$\Rightarrow |z_1 z_2| = |z_1||z_2|$ [1]

3 (i) $|z - 3 - 3i| = 3$ [1], [1]

(ii) $\arg(z - 3 - 3i) = \dfrac{\pi}{3}$ [1], [1]

(iii) $\dfrac{1}{2}r^2\theta = \dfrac{3\pi}{8}$ [1]

$r = 3 \Rightarrow \theta = \dfrac{\pi}{12}$ [1]

$\Rightarrow \arg(z - 3 - 3i) = \dfrac{\pi}{3} + \dfrac{\pi}{12} = \dfrac{5\pi}{12}$ [1]

So the half line has equation

$\arg(z - 3 - 3i) = \dfrac{5\pi}{12}$. [1]

4 (i) $5x - y - 25 = 0$ [1], [1]

(ii)

$\begin{pmatrix} 5 \\ -1 \\ 0 \end{pmatrix} \cdot \begin{pmatrix} 4 \\ -3 \\ 1 \end{pmatrix} = \left|\begin{pmatrix} 5 \\ -1 \\ 0 \end{pmatrix}\right| \left|\begin{pmatrix} 4 \\ -3 \\ 1 \end{pmatrix}\right| \cos\theta$ [1]

$\Rightarrow 23 = \sqrt{26}\sqrt{26}\cos\theta$

$\Rightarrow \dfrac{23}{26} = \cos\theta \Rightarrow \theta = 27.8°$ [1], [1]

(iii) $5 \times 5 - 0 - 25 = 0$ and [1], [1]

$4 \times 5 - 3 \times 0 + (-17) - 3 = 0$

5 (i) [1]

(ii) $V = \pi \displaystyle\int_0^{2\pi} (1 - \cos x)^2 \, dx$ [1]

$= \pi \displaystyle\int_0^{2\pi} (1 - 2\cos x + \cos^2 x) \, dx$

$= \pi \displaystyle\int_0^{2\pi} \left(1 - 2\cos x + \dfrac{1}{2}(1 + \cos 2x)\right) \cdot dx$ [1]

$= \pi \left[x - 2\sin x + \dfrac{1}{2}(x + \dfrac{1}{2}\sin 2x)\right]_0^{2\pi}$ [1]

$= 3\pi^2$ [1]

6 (i) $(0, 1, 2) \Rightarrow n_2 + 2n_3 + d = 0$

$(1, 1, 2) \Rightarrow n_1 + n_2 + 2n_3 + d = 0$

$(3, 3, 3) \Rightarrow 3n_1 + 3n_2 + 3n_3 + d = 0$ [1]

so $n_1 = 0$ [1]

and $n_2 + 2n_3 + d = 0$, $3n_2 + 3n_3 + d = 0$

$\Rightarrow n_2 = \dfrac{1}{3}d$, $n_3 = -\dfrac{2}{3}d$ [1]

So Π is (any non-zero multiple of)

$y - 2z + 3 = 0$ Or any equivalent method

(ii) $\begin{pmatrix} 3 \\ 0 \\ 4 \end{pmatrix}$ (or any non-zero multiple of this) [1]

(iii) Line: $x = 1 + 3t$, $y = 2$, $z = 3 + 4t$

Plane: $y - 2z + 3 = 0$

So at intersection: $2 - 2(3 + 4t) + 3 = 0$ [1]

$\Rightarrow t = -\dfrac{1}{8}$ [1]

$\Rightarrow \left(\dfrac{5}{8}, 2, 2\dfrac{1}{2}\right)$ [1]

(iv) $\lambda = \dfrac{\begin{pmatrix} 3-1 \\ 3-2 \\ 3-3 \end{pmatrix}\begin{pmatrix} 3 \\ 0 \\ 4 \end{pmatrix}}{\begin{pmatrix} 3 \\ 0 \\ 4 \end{pmatrix}\begin{pmatrix} 3 \\ 0 \\ 4 \end{pmatrix}} = \dfrac{6}{25}$ [1]

Distance $= \left\| \begin{pmatrix} 1-3 \\ 2-3 \\ 3-3 \end{pmatrix} + \lambda \begin{pmatrix} 3 \\ 0 \\ 4 \end{pmatrix} \right\|$

$= \left\| \begin{pmatrix} -1.28 \\ -1 \\ 0.96 \end{pmatrix} \right\|$ [1]

$= \sqrt{3.56} = 1.89$ [1]

Or any equivalent method
The shortest distance from the point P, with position vector $\mathbf{p}$, to the line $\mathbf{r} = \mathbf{a} + \lambda\mathbf{b}$ is $|(\mathbf{a} - \mathbf{p}) + \lambda\mathbf{b}|$, where $\lambda = \dfrac{(\mathbf{p} - \mathbf{a}).\mathbf{b}}{\mathbf{b}.\mathbf{b}}$.

7 (i) $\begin{pmatrix} 20 & 0 & 1 \\ -20 & 0 & 1 \\ 0 & -20 & -1 \\ 0 & 20 & -1 \end{pmatrix}\begin{pmatrix} 5 \\ 17 \\ 20 \end{pmatrix}$

$= \begin{pmatrix} k \\ l \\ m \\ n \end{pmatrix} = \begin{pmatrix} 120 \\ -80 \\ -360 \\ 320 \end{pmatrix}$

$\Rightarrow k = 120, l = -80, m = -360,$

$n = 320$ [1], [1]

There is no need to use matrices, you could simply substitute the coordinates of the vertex, which must be in all four planes, into the equation of each plane to find k, l, m and n.

(ii) Because the summit is directly above the centre of the square base, each face makes the same angle to the vertical.

$\begin{pmatrix} 20 \\ 0 \\ 1 \end{pmatrix}.\begin{pmatrix} 0 \\ 0 \\ 1 \end{pmatrix} =$

$= \left\| \begin{pmatrix} 20 \\ 0 \\ 1 \end{pmatrix} \right\| \left\| \begin{pmatrix} 0 \\ 0 \\ 1 \end{pmatrix} \right\| \cos\theta$ [1]

$\Rightarrow 1 = \sqrt{401}\cos\theta$

$\Rightarrow \theta = \cos^{-1}\left(\dfrac{1}{\sqrt{401}}\right) = 87.1°$ [1], [1]

So each face makes an angle of 2.9° to the vertical. [1]

(iii) The summit of the skyscraper is where any three of the four triangular faces intersect. [1]

$\begin{array}{l} 25x + z = 150 \\ -25x + z = -100 \end{array}$ Adding gives $z = 25$ [1]

$\begin{array}{l} 25y - z = 250 \\ 25y + z = 300 \end{array}$ Adding gives $y = 11$

$\begin{array}{l} 25x + z = 150 \\ 25y - z = 250 \end{array}$ Substituting in $y = 11$ and adding gives $x = 5$

So the coordinates of the summit are $(5, 11, 25)$ [2]

(iv) A very tall skyscraper might be 300 m high. [1]
The z-coordinate of the summit is 25, suggesting each unit might be $\dfrac{300}{25} = 12$ m. [1], [1]

Other answers, suitably justified, are acceptable.

8 (i) $a = -10, b = 14, c = -160$, or

$a = -\dfrac{19}{2}, b = -\dfrac{17}{2}, c = 7$ [1], [1], [1]

(ii) $a = -10, b = 14$, c is any number other than -160

or $a = -\dfrac{19}{2}, b = -\dfrac{17}{2}$, c is any number other than 7 [2]

Answers

(iii) If the planes meet at a single point, the point (x, y, z) where the three planes meet can be represented by the matrix equation

$$\begin{pmatrix} 5 & -7 & 1 \\ 1 & -13 & -2 \\ 19 & 17 & -4 \end{pmatrix} \begin{pmatrix} x \\ y \\ z \end{pmatrix}$$

$$= \begin{pmatrix} 80 \\ -2 \\ -14 \end{pmatrix} \Rightarrow \begin{pmatrix} x \\ y \\ z \end{pmatrix}$$

$$= \begin{pmatrix} 5 & -7 & 1 \\ 1 & -13 & -2 \\ 19 & 17 & -4 \end{pmatrix}^{-1} \begin{pmatrix} 80 \\ -2 \\ -14 \end{pmatrix} \quad [1], [1]$$

Using a calculator,

$$\begin{pmatrix} 5 & -7 & 1 \\ 1 & -13 & -2 \\ 19 & 17 & -4 \end{pmatrix}^{-1}$$

$$= \frac{1}{932}\begin{pmatrix} 86 & -11 & 27 \\ -34 & -39 & 11 \\ 264 & -218 & -58 \end{pmatrix}.$$

Since this inverse matrix exists, the planes must meet at a single point. [1]

$$\begin{pmatrix} x \\ y \\ z \end{pmatrix} = \frac{1}{932}\begin{pmatrix} 86 & -11 & 27 \\ -34 & -39 & 11 \\ 264 & -218 & -58 \end{pmatrix}\begin{pmatrix} 80 \\ -2 \\ -14 \end{pmatrix} \quad [1]$$

$$= \frac{1}{932}\begin{pmatrix} 6524 \\ -2796 \\ 22368 \end{pmatrix} = \begin{pmatrix} 7 \\ -3 \\ 24 \end{pmatrix} \quad [1]$$

So the planes meet at $(7, -3, 24)$ [1]

An introduction to radians
Exercise (Page 202)

1 (i) $\dfrac{\pi}{3}$ (ii) $\dfrac{\pi}{4}$ (iii) $\dfrac{5\pi}{6}$

(iv) $\dfrac{10\pi}{9}$ (v) 0.775^{C} (3 s.f.) (vi) $\dfrac{9\pi}{4}$

(vii) $\dfrac{3\pi}{2}$ (viii) 1.73^{C} (3 s.f.) or $\dfrac{11\pi}{20}$

(ix) $\dfrac{5\pi}{3}$ (x) 4π

(xi) $\dfrac{\pi}{12}$ (xii) $\dfrac{\pi}{60}$ or 0.0524^{C} (3 s.f.)

2 (i) 20° (ii) 24°

(iii) 229° (3 s.f.) (iv) 300°

(v) 25.7° (3 s.f.) (vi) 9°

(vii) 103° (3 s.f.) (viii) 220°

(ix) 630° (x) 900°

(xi) 405° (xii) 255°

The identities $(\theta \pm \phi)$ and $(\theta \pm \phi)$
Exercise (Page 204)

1 (i) $\dfrac{1 + \sqrt{3}}{2\sqrt{2}}$ (ii) $\dfrac{1 + \sqrt{3}}{2\sqrt{2}}$

(iii) $\dfrac{1 + \sqrt{3}}{2\sqrt{2}}$ (iv) $\dfrac{\sqrt{3} - 1}{\sqrt{2}}$

2 (i) $\dfrac{1}{2}$ (ii) 1

(iii) $\cos 4\theta$ (iv) $\dfrac{\sqrt{3}}{2}$

(v) $\dfrac{\sqrt{3} - 1}{\sqrt{2}}$ (vi) $\dfrac{1}{2}$

3 (i) $\dfrac{1}{\sqrt{2}}(\sin \theta + \cos \theta)$

(ii) $\dfrac{\sqrt{3}}{2}\cos 2\theta + \dfrac{1}{2}\sin 2\theta$

(iii) $\dfrac{\sqrt{3}}{2}\sin \theta - \dfrac{1}{2}\cos \theta$

(iv) $\dfrac{1}{2}\cos 3\theta - \dfrac{\sqrt{3}}{2}\sin 3\theta$

Index

Page numbers followed by *f* indicate figures.

A

addition
    complex numbers, 37
    formulae, 203
    matrices, 3
angle between vectors, 147–54, 148*f*, 150*f*, 151*f*
Argand diagram, 97–100, 97–100*f*
    complex numbers, 97–8, 97–8*f*
    loci in *see* loci in Argand diagram
    modulus–argument form of complex numbers, 103, 103*f*
associativity, 4
    matrix multiplication, 9–10

B

Bombelli, Rafael, 38

C

Cardano, Gerolamo, 38
Cartesian equation of a line, vector equation, 157–8
commutability, 4
    matrix multiplication, 9–10
complex conjugates, 40–1
complex numbers, 35–47
    Argand diagram, 97–8, 97–100*f*
        see also loci in Argand diagram
    Argand diagram in modulus–argument form of, 103, 103*f*
    definition, 36–7
    difference representation, 44–5
    division, 40–3
    equality of, 38–40
    geometric representation, 43–7, 43*f*, 44*f*
    modulus–argument form *see* modulus–argument form of complex numbers
    notation, 37
    sum representation, 44–5, 45*f*
    working with, 37–8
complex roots, polynomials, 48, 61–5, 62*f*, 63*f*
compound angle formulae, 203
conjectures, 75
convergent sequence, 67
$\cos[\theta \pm \phi]$, 203–4
counter-examples, 75
cubic equations, polynomials, 54–8

D

decreasing sequence, 67
Descartes, René, 38
determinants of matrices *see* matrices, determinants of

difference representation, complex numbers, 44–5
distances, 186–95
    from point to line, 186–7, 186*f*
    from point to plane, 187–8, 188*f*
    between skew lines, 190–2, 190*f*
    between two parallel lines, 189–90, 189*f*
division
    complex numbers, 40–3
    modulus–argument form of complex numbers, 105–8, 105*f*, 107*f*
dot product, 149

E

elements, matrices, 2
equality of complex numbers, 38–40
equating real and imaginary parts, 38
equation of plane, 166–75, 166*f*, 167*f*, 169*f*
    notation, 170–1, 170*f*
Euler, Leonhard, 38

F

formulae derivation, polynomials quartic equations, 58–60

G

geometrical interpretation in two dimensions, 141–5, 142*f*, 143*f*
geometric representation of complex numbers, 43–7, 43*f*, 44*f*

I

identity matrices, 3, 129–30
image, transformations, 13
imaginary parts, 38
increasing sequence, 67
induction, proof by
    sum of sequence, 79–83
    sum of series, 76–9
inductive definition of sequence, 67
integers, 35*f*, 36
integration, 86–95
intersection of planes, 175–82
    arrangement of three planes, 177–80, 179–80*f*
    geometric arrangement of three planes, 175–6, 175–6*f*
    intersection of three planes, 176–7
invariance, 28–33
    lines, 29–33, 29*f*, 30*f*, 31*f*
    points, 28–9
inverse of a product of matrices, 132–4, 133*f*
inverse of matrices, 129–34
    3 × 3 matrix, 135–40
irrational numbers, 36

L

Leidniz, Gottfried, 38
lines
    angle between a plane and, 183
    distance between two parallel lines, 189–90, 189*f*
    distances between skew lines, 190–2, 190*f*
    distance to a point, 186–7, 186*f*
    invariance, 29–33, 29*f*, 30*f*, 31*f*
    invariant points of, 28
    point of intersection with plane, 182
    vector equation of *see* lines, vector equation of
lines, vector equation of, 154–66
    angle between two lines, 161–2, 161*f*
    Cartesian equation of a line, 157–8
    intersection of straight line in three dimensions, 159–61
    intersection of straight line in two dimensions, 158–9
    lines in three dimensions, 157
    lines in two dimensions, 154–6, 155*f*
loci in Argand diagram, 109–20*f*, 109–21
    form $\arg(z - a) = \theta$, 112–14, *112–14f*
    form $[z - a] = r$, 109–12, 109–12*f*
    form $|z - a| = |z - b|$, 114–16, 115–16*f*

M

Mandlebrot set, 96*f*
mapping, transformations, 13
mathematical induction, 75–9
    definition, 75
    proof by *see* induction, proof by
matrices, 2–6, 5*f*
    definition, 2
    determinants of *see* matrices, determinants of
    discussion point, 1*f*
    elements, 2
    inverse of, 129–34
    inverse of a product of, 132–4, 133*f*
    inverse of 3 × 3 matrix, 135–40
    multiplication *see* matrix multiplication
    non-conformable, 3
    non-singular inverse, 131
    order, 2
    simultaneous equation solutions, 141–8
    transformation, 126
    transformation representation, 14–19
    types, 2–3
    working with, 3
matrices, determinants of, 123–9, 123*f*, 124*f*

Index